T0155574

Design Automation Techniques for Approximation Circuits

Arun Chandrasekharan • Daniel Große
Rolf Drechsler

Design Automation Techniques for Approximation Circuits

Verification, Synthesis and Test

 Springer

Arun Chandrasekharan
OneSpin Solutions GmbH
Munich, Germany

Daniel Große
University of Bremen and DFKI GmbH
Bremen, Germany

Rolf Drechsler
University of Bremen and DFKI GmbH
Bremen, Germany

ISBN 978-3-030-07550-7   ISBN 978-3-319-98965-5   (eBook)
https://doi.org/10.1007/978-3-319-98965-5

This Springer imprint is published by the registered company Springer Nature Switzerland AG
The registered company address is: Gewerbestrasse 11, 6330 Cham, Switzerland

To Keerthana,
Nanno
and
Zehra

Preface

APPROXIMATE COMPUTING is a novel design paradigm to address the performance
and energy efficiency needed for future computing systems. It is based on the
observation that many applications compute their results more accurately than
needed, wasting precious computational resources. Compounded to this problem,
dark silicon and device scaling limits in the hardware design severely undermine
the growing demand for computational power. Approximate computing tackles this
by deliberately introducing controlled inaccuracies in the hardware and software to
improve performance. There is a huge set of applications from multi-media, data
analytics, deep learning, etc. that can make a significant difference in performance
and energy efficiency using approximate computing. However, despite its potential,
this novel computational paradigm is in its infancy. This is due to the lack of efficient
design automation techniques that are synergetic to approximate computing. Our
book bridges this gap. We explain algorithms and methodologies from automated
synthesis to verification and test of an approximate computing system. All the
algorithms explained in this book are implemented and thoroughly evaluated on a
wide range of benchmarks and use cases. Our methodologies are efficient, scalable,
and significantly advance the state of the art of the approximate system design.

Acknowledgments

First and foremost, we would like to thank the members of the research group for computer architecture at the University of Bremen. We deeply appreciate the continuous support, the inspiring discussions, and the stimulating environment provided. Next, we would like to thank all coauthors of the papers which formed the starting point for this book: Mathias Soeken, Ulrich Kühne, and Stephan Eggersglüß. This book would not have been possible without their academic knowledge and valuable insight. Our sincere thanks also go to Kenneth Schmitz, Saman Fröhlich, and Arighna Deb for numerous discussions and successful collaborations.

Munich, Germany   Arun Chandrasekharan
Bremen, Germany   Daniel Große
Bremen, Germany   Rolf Drechsler
July 2018

Contents

List of Algorithms

List of Figures

List of Tables

Chapter 1
Introduction

APPROXIMATE COMPUTING is an emerging design paradigm to address the performance and energy efficiency needed for the future computing systems. Conventional strategies to improve the hardware performance such as device scaling have already reached its limits. Current device technologies such as 10 nm are already reported to have significant secondary effects such as quantum tunneling. On the energy front, dark silicon and the power density is a serious challenge and limiting factor for several contemporary IC design flows. It is imperative that the current state-of-the-art techniques are inadequate to meet the growing demands of the computational power. Approximate computing can potentially address these challenges. It refers to hardware and software techniques where the implementation is allowed to differ from the specification, but within an acceptable range. The approximate computing paradigm delivers *performance* at the cost of *accuracy*. The key idea is to trade off correct computations against energy or performance. At a first glance, one might think that this approach is not a good idea. But it has become evident that there is a huge set of applications which can tolerate errors. Applications such as multi-media processing and compressing, voice recognition, web search, or deep learning are just a few examples. However, despite its huge potential, approximate computing is not a mainstream technology yet. This is due to the lack of reliable and efficient design automation techniques for the design and implementation of an approximate computing system. This book bridges this gap. Our work addresses the important facets of approximate computing hardware design—from formal verification and error guarantees to synthesis and test of approximation systems. We provide algorithms and methodologies based on classical formal verification, synthesis, and test techniques for an approximate computing IC design flow. Further, towards the end, a novel hardware architecture is presented for cross-layer approximate computing. Based on the contributions, we advance the current state-of-the-art of the approximate hardware design.

Several applications spend a huge amount of energy to guarantee correctness. However, correctness is not always required due to some inherent characteristics

© Springer Nature Switzerland AG 2019

A. Chandrasekharan et al., *Design Automation Techniques*
for Approximation Circuits, https://doi.org/10.1007/978-3-319-98965-5_1

of the application. For example, recognition, data mining, and synthesis (RMS) applications are highly computationally intensive, but use probabilistic algorithms. Here, the accuracy of the results can be improved over successive iterations or by using a large number of input samples. Certain applications such as video, audio, and image processing have perceptive resilience due to limited human skills in understanding and recognizing details. Certain other applications such as database and web search do not necessarily have the concept of a unique and correct answer. Such applications require large amount of resources for exact computations. However, all these applications can provide huge gains in performance and energy when *bounded* and *controlled* approximations are allowed, while still maintaining the acceptable accuracy of the results. This key observation is the foundation of approximate computing paradigm. For several applications, time and resources spend on algorithms that can be approximated can even go up to 90% of the total computational requirements [CCRR13]. Needless to say, such applications benefit immensely using approximate computing techniques.

Certainly, several questions arise when rethinking the design process under the concept of approximate computing:

1. What errors are acceptable for a concrete application?
2. How to design and synthesize approximate circuits?
3. How to perform functional verification?
4. What is the impact of approximations in production test?
5. How can approximate circuits be diagnosed?
    etc...

All of these are major questions. This book proposes design automation methodologies for the synthesis, verification, and test of the approximate computing hardware. Thus, we directly address the 2nd, 3rd, and 4th question.

For answering the first question, different error metrics have been proposed. Essentially, they measure the approximation error by comparing the output of the original circuit against the output of the approximation circuit. Typical metrics are error-rate, worst-case error, and bit-flip error. The chosen metric depends highly on the application.

On the design side (second question), this book focuses on *functional* approximations, i.e., a slightly different function is realized (in comparison to the original one) resulting in a more efficient implementation. The primary focus of this book is on hardware approximation systems. Two main directions of functional approximation for a hardware can be distinguished: (1) for a given design, an approximation circuit is created manually; most of the research has been done here. This includes, for example, approximate adders [SAHH15, GMP+11, KK12, MHGO12] and approximate multipliers [KGE11]. However, since this procedure has strong limitations in making the potential of approximation widely available, research started on (2) design automation methods to derive the approximated components from a golden design automatically.

Different approximation synthesis approaches have been proposed in the literature. They range from the reduction of sum-of-product implementations [SG10], redundancy propagation [SG11], and don't care based simplification (SALSA) [VSK$^+$12] to dedicated three-level circuit construction heuristics [BC14]. Recently, the synthesis framework ASLAN [RRV$^+$14] has been presented, which extends SALSA and is able to synthesize approximate sequential circuits. ASLAN uses formal verification techniques to ensure quality constraints given in the form of a user-specified *Quality Evaluation Circuit* (QEC). However, the QEC has to be constructed by the user similar to a test bench, which is a design problem by itself. In addition, constructing a circuit to formulate the approximation error metrics requires detailed understanding of formal property checking (liveness and safety properties) and verification techniques. Further, some error metrics such as error-rate cannot be expressed in terms of Boolean functions efficiently since these require counting in the solution space, which is a #SAT problem (i.e., counting the number of solutions). Moreover, the error metrics used in these approaches are rather restricted (e.g., [VSK$^+$12] uses worst-case error and a very closely related relative-error as metrics) and how to trade off a stricter requirement in one metric wrt. to a relaxed requirement in another has not been considered. This is important when the error metrics are unrelated to each other. Therefore, the current approximation synthesis system techniques are severely limited and inadequate. This book introduces new algorithms and methodologies for the approximation synthesis problem. The evaluations of these methodologies are carried out on a wide range circuits. The proposed algorithms are effective and scalable and come with a formal guarantee on the error metrics.

Precisely computing error metrics in an approximate computing hardware is a hard problem, but it is inevitable when aiming for high quality results or when trading off candidates in design space exploration. Since the very idea of approximate computing relies on controlled insertion of errors, the resulting behavior has to be carefully studied and verified. This addresses the third question: "how to perform functional verification?". In the past, approaches based on simulation and statistical analysis have been proposed for approximate computing [VARR11, XMK16]. However, all such approaches are dependent on a particular error model and a probabilistic distribution. Hence, very few can provide formal guarantees. Furthermore, in sequential circuits, errors can accumulate and become critical over time. In the end, the error behavior of an approximated sequential circuit is distinctly different from that of an approximated combinational circuit. We have developed algorithms and methodologies that can determine and prove the limits of approximation errors, both in combinational and sequential systems. As our results show, formal verification for error guarantees is a must for approximate computing to be mainstream.

The next question concerns with the impact of approximation in production test after the integrated circuit manufacturing process. This book investigates this aspect and proposes an approximation-aware test methodology to improve the production yield. To the best of our knowledge, this is the first approach considering the impact of design level approximations in post-production test. Our results show that there is a significant potential for yield improvement using the proposed approximation-aware test methodology.

In this book, we propose the design automation techniques for the synthesis, verification, and test of an approximate hardware design. Nevertheless, the approximate computing paradigm is not limited to the underlying hardware alone. Naturally, the software can also perform standalone approximations without involving the hardware. However, in several cases, a cross-layer approximate computing strategy involving both the hardware and the software level approximations is the most beneficial for performance [VCC$^+$13, YPS$^+$15]. This brings out the best of both worlds. We dedicate the final chapter of this book to such cross-layer approximation architectures.

In a cross-layer approximate computing scheme, the software and the hardware work in tandem, reinforcing each ones capabilities. To achieve such a system, certain architectural level approximation features are needed. The software actively utilizes the underlying hardware approximations and dynamically adapts to the changing requirements. There are several system architectures proposed ranging from those employing neural networks to dedicated approximation processors [YPS$^+$15, SLJ$^+$13, VCC$^+$13, CWK$^+$15]. The hardware is designed wrt. an architectural specification on the respective error criteria. Further, the software that runs the end application is aware of such hardware features and actively utilizes them. The final chapter of this book provides the details of an important microprocessor architecture we have developed for approximate computing. This architecture called ProACt can do cross-layer approximations spanning hardware and software. ProACt stands for *Processor for On-demand Approximate Computing*. The ProACt processor can dynamically adapt to the changing external computational demands.

Each chapter in this book focuses on one main contribution towards the realization of a high performance approximate computing systems, answering several of the major questions outlined earlier. An approximate computing IC design flow is not radically different from a conventional one. However, there are some important modifications and additional steps needed to incorporate the principles of approximate computing. An overview of the approximate computing IC design flow along with the important contributions in this book is given next.

1.1 Approximate Computing IC Design Flow

Figure 1.1 shows an overview of a typical approximate computing IC design flow. It starts with the architectural specifications and ends with the final IC ready to be shipped to the customer. The colored boxes in the block diagram indicate the important contributions made in this work.

The architectural specifications (or simply the specs) for an IC targeted for approximate computing will usually have error requirements. Sometimes, the architectural stage is divided into broad specs and micro-architectural level specs. Micro-architectural specs tend to be detailed and contain information about the block level features of the product, e.g., in case of a *System-on-Chip* (SoC), the number of pipeline stages for the processor block. In this case, the broad specs

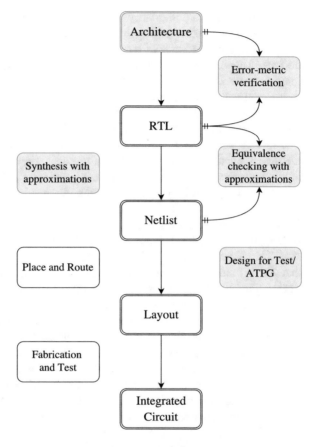

Fig. 1.1 Design flow for approximate computing IC (colored boxes are the book contributions)

may only detail the address-bus width of the processor and the amount of memory to be supported. In any case, it is important to consider the error criteria also. As mentioned before we introduce an on-demand approximation capable processor in this book. The processor called ProACt has several micro-architectural level features that make it an excellent candidate for cross-layer approximate computing. The implemented hardware prototype of ProACt is a 64-bit, 5-stage pipeline processor incorporating state-of-the-art system architecture features such as L1, L2 cache, and DMA for high throughput.

The next step is to implement or integrate the individual block designs together. This step is shown as "RTL" (stands for *Register-Transfer-Level*, a widely used design abstraction to model the hardware) in Fig. 1.1 and is mostly manual. Designers write the code in a high level language such as VHDL or Verilog with several hardware IPs. For approximate computing, there has been intensive research in this area, with researchers reporting approximation tuned circuits. Refer to [XMK16, VCRR15] for an overview of such works. However, a critical aspect that must be ensured at this stage is to verify the design for the intended error specification. This is particularly important when the approximated component is used together with other non-approximated ones or in sequential circuits where the impact of approximation is different from the combinational circuits. Many of the arithmetic approximation circuits reported and the associated literature do not offer this full scale analysis. We have devised algorithms and methodologies based on formal verification to aid this analysis.

The design is further synthesized to a *netlist*—the optimized structural representation targeted to a production technology. Here, there are two important aspects. The first aspect is the automated synthesis that takes in the approximation error metrics along with the RTL and writes out an optimized netlist. The second one is the equivalence checking after this approximation synthesis. It is important in any synthesis approach to verify the results. As mentioned before, this book makes very important contributions to both: synthesis and equivalence checking in the presence of approximations. After synthesis, the quality of the approximated circuits is formally evaluated in a separate subsequent stage, independent of the synthesis step. This is important in several aspects. First and foremost it gives a guarantee that the synthesis tool is free of programmer errors. This also aids in design exploration, to compare the results with other schemes such as a pure automated approximation synthesis solution vs. an architecturally approximated solution. Besides, the specifications on the error metrics are the limiting conditions provided as input to the synthesis tool and the tool would have achieved a lesser optimal results. This has to be seen in the context of approximate computing applications. Each application has a different tolerance on a specific error criteria. Moreover, several error metrics are orthogonal to each other, i.e., a higher requirement on one metric does not imply a similar requirement on another. An example is the bit-flip error and worst-case error. Hence, it is important to formally verify the achieved result after the approximation synthesis.

Once the netlist is obtained, the remaining implementation steps are similar to conventional IC design. The netlist is taken through further steps such as placement, clock-tree synthesis, and routing and finally a layout is generated.[1] This layout is send to the fabrication house and the fabricated IC is tested for manufacturing defects. One important activity parallel to implementation is *Design For Test* (DFT). The post-production test after fabrication involves testing the IC using *test patterns*. These test patterns are generated in the DFT stage. This step is called *Automated Test Pattern Generation* (ATPG). Note that the test patterns are different from functional vectors used in functional simulation. Test patterns are rather engineered to detect the manufacturing defects and sort out the defective chip out of the lot. However, as shown later in this book, the knowledge of functional approximation can make a difference in the test pattern generation. The final yield for the manufacturing

---

[1]In reality, Fig. 1.1 is an oversimplification of the actual IC design flow. There could be several iterations and sign-off checks from netlist to layout, before sending the final layout data to fabrication.

process from the fab house depends on the testing step. An approximation-aware test generation has the potential to significantly improve the final yield.

An outline of the book is provided next. This will facilitate to understand the different steps in detail, and about the book organization. The CAD tools for all the algorithms and methodologies presented in this book are implemented and verified. Most of them are publicly available also. The *aXc* software framework contains most of these techniques. Further details on *aXc* tool and other software and hardware frameworks are provided towards the end of this chapter.

1.2 Outline

The contributions in this book are organized in Chaps. 3–7. Each chapter discusses an important step in the realization of approximate computing.

Chapter 2 explains the necessary preliminary materials needed for the sound understanding of the algorithms and methodologies described. This chapter includes the basics of data structures, formal verification techniques, and an overview of the important error metrics used in approximate computing.

As mentioned before, there exists a large body of literature for ad-hoc designed or architected approximation circuits (see [XMK16] for an overview). It is imperative to check whether these circuits indeed conform to the bounds of the error behavior specified or not. Formal verification and in particular logical equivalence checking are important in several stages of chip design. It is very common to perform an equivalence check after most of the steps that modify the logic and structure of the design. For example, while designing a chip, logical equivalence checking is conducted after synthesis (behavioral description to technology dependent structural transformation stage), test insertion (scan-stitching, DFT architecting), clock-tree synthesis, and several post-route and pre-route timing optimization steps. It is fair to say that formal verification is one of the most important sign-off checks in today's IC design. Formal verification has to be applied in the same rigor for approximate computing designs also. Hence, this book has two dedicated chapters on the formal verification techniques we have developed for approximate computing circuits—Chap. 3 for combinational circuit verification and Chap. 4 for sequential approximated circuits. Note that there is an important distinction here compared to classical methods. The formal verification as applied to approximate computing circuits must guarantee the error bounds. Thus, the conventional definition of equivalence checking does not strictly hold here. Chapter 2 details these aspects on error metrics.

Chapter 5 concentrates on the automated synthesis techniques for approximate computing. Conventional synthesis takes in a RTL description of the design and converts it into a structural netlist mapped to a technology library targeted to a foundry. Approximate synthesis reads in the same RTL description along with a user-specified error metric. Further, approximate synthesis generates a netlist with approximations that conforms to the user-specified error metrics with performance optimizations in area, timing, and power.

An important aspect in IC design is post-production test. Approximate computing circuits offer several unique advantages in terms of test generation. We have developed techniques to aid the testing of approximate circuits which is provided in Chap. 6.

The next chapter, Chap. 7 is related to the architecture of approximate computing processors. We present an open source state-of-the-art high performance approximate computing processor that can do on the fly approximations. This chapter is different from the earlier chapters in several ways. Here, we examine the important architectural specification that can be used for cross-layer approximate computing. Cross-layer approximate computing involves the hardware and the software working together to achieve performance benefits through approximation. Instead of a CAD tool or methodology, this chapter details the on-demand approximation processor ProACt—the implementation, *Instruction Set Architecture* (ISA), FPGA prototyping, compilers, and approximation system libraries.

The concluding remarks for this work are provided in the final chapter, Chap. 8. The *aXc* framework has the implementation for most of the algorithms and methodologies explained in this book. Details on this software framework are provided next.

1.3   AxC Software Framework and Related Tools

The algorithms presented in Chaps. 3–6 are implemented in the *aXc* framework. *aXc* has a dedicated command line interface where the user can input different commands. These commands include, e.g., those to do synthesis, verification, read/write different design formats, etc. All the commands have a *help* option invoked using the $-h$ or $-help$ option (e.g., $report\_error - h$ to show the help for the $report\_error$ command). Further, the tool can read and write several design formats including Verilog. See the tool documentation for a complete list of the supported commands and features. *aXc* is publicly distributed with Eclipse Public License (EPL 1.0)[2] and is available here[3]:

- https://gitlab.com/arunc/axekit

---

[2]Note: *aXc* software framework uses several other third party packages and software components. All the third party packages come with own licensing requirements. See *aXc* documentation for a full list of such packages and related licenses.

[3]Note: the test algorithms provided in Chap. 6 are also implemented in *aXc*. However, these are proprietary and not part of the publicly available *aXc* software distribution.

An example for a typical *aXc* invocation is shown below. The snippet is self-explanatory. It shows how to read a reference design in Verilog (called golden design in *aXc*) and an approximated version of it, and further proceed with approximation error reports.

```
#----------------------------------------------------
aXc> version
      AxC build: npn-285fbb3
aXc> read_verilog golden.v --top_module golden
aXc> read_verilog approx.v --top_module approx
aXc> report_error --all --report error_metrics.rpt
      [i] worst_case_error    = 64
      [i] average_case_error  = 7.5
      [i] bit_flip_error      = 5
      [i] error_rate          = 96   ( 18.75  % )
aXc> quit
#----------------------------------------------------
```

As mentioned, Chap. 7 explains the details of the ProACt processor. All the related research materials are publicly available in the following repositories:

- https://gitlab.com/arunc/proact-processor : ProACt processor design
- https://gitlab.com/arunc/proact-zedboard : Reference hardware prototype implementation
- https://gitlab.com/arunc/proact-apps : Application development using ProACt and approximation libraries

Further details on approximate computing tools can be found in other repositories such as Yise (https://gitlab.com/arunc/yise) and Maniac (http://www.informatik.uni-bremen.de/agra/ger/maniac.php).

Chapter 2
Preliminaries

This chapter explains the necessary concepts needed for the sound understanding of the approximate computing techniques and algorithms discussed in the later chapters. Concise and scalable data structures that can represent the logic of a circuit is crucial for the success of *Electronic Design Automation* (EDA). Several EDA algorithms to be introduced in the subsequent chapters directly benefit from the underlying data structure. Hence, this chapter starts with an overview of the relevant data structures such as *Binary Decision Diagrams* (BDDs) and *And-Inverter Graphs* (AIGs). Several algorithms presented in this book rely heavily on Boolean Satisfiability (SAT). The verification, synthesis, and test techniques for approximate computing are based on SAT. The relevant details about the combinatorial satisfiability problem along with the variants of the classical SAT techniques useful for approximate computing are introduced in the second part. Further, important concepts on the post-production test and *Automated Test Pattern Generation* (ATPG) are outlined in the next section. This is required for Chap. 6 on test for approximate computing. A discussion on the different error metrics used in approximate computing forms the last part of this chapter. Approximate computing applications express their tolerance using these error metrics. Before getting into the relevant details on these topics, the common notations and conventions used in this book are outlined next.

2.1 General Notation and Conventions

In this book, $\mathbb{B}$ represent a *Boolean domain* consisting of the set of binary values $\{0, 1\}$. A *Boolean variable* x is a variable that takes values in some Boolean domain; $x \in \mathbb{B}$. $f : \mathbb{B}^n \to \mathbb{B}^m$ is a *Boolean function* with n primary inputs and m primary outputs, i.e., $f(x_0, \ldots, x_{n-1}) = (f_{m-1}, \ldots, f_0)$. The domain and the co-domain (alternately called *range*) of f is Boolean. There are 2^n input combinations for each

© Springer Nature Switzerland AG 2019
A. Chandrasekharan et al., *Design Automation Techniques for Approximation Circuits*, https://doi.org/10.1007/978-3-319-98965-5_2

of the m outputs of the Boolean function f. The logic primitives, Boolean AND (conjunction operation) is shown as $\wedge$, Boolean OR (disjunction operation) using $\vee$, and the exclusive-or operation using $\oplus$. We use an over-bar (e.g., $\bar{x}$) or $!x$ for the negation operation. Further, the $\wedge$ symbol is omitted if it is understood from the context, i.e., the Boolean expression $x_1 \wedge x_2$ may be simply written as $x_1 x_2$. Other notations are introduced in the respective contexts when needed.

2.2 Data Structures: Boolean Networks

Modern state-of-the-art electronic systems experience a rapid growth in terms of complexity. These days, designs with multi-million logic elements are quite common. Therefore, the EDA tools have to deal with huge volume of design data. This calls for a scalable and efficient underlying data structure that can be easily manipulated. Here, we deal with the data structures of digital approximation circuits that abide by the rules of *Boolean algebra*. Hence, the data structures that are of primary interest are *Boolean networks*. Throughout the history of EDA, several forms of Boolean networks have been used such as *Sum-of-Products* (SOP), *Binary Decision Diagrams* (BDD), and *And-Inverter Graphs* (AIGs). These concise data structures are integral to several algorithms used inside the EDA tools.

A *Boolean network* is a *Directed Acyclic Graph* (DAG) where nodes (vertices) represent a logic primitive (Boolean gate) or Primary Inputs/Primary Outputs (PIs/POs), and edges represent wires that form the interconnection among the primitives. Note that in a general Boolean network representation nodes/edges can have polarity showing inversion. Since these networks form the backbone of EDA tools, it is imperative that any digital logic should be able to be represented as a Boolean network. Such a representation is called *Universal representation* or *Universal gate*. See [Knu11] for detailed discussions on Boolean algebra and its postulates.

A Boolean network can be *homogeneous* or *non-homogeneous*. Homogeneous networks are composed of only one kind of Boolean primitive such as a Boolean AND gate. On the contrary, a non-homogeneous network contains different kinds of Boolean gates. Here, for efficiency the functionality of each Boolean primitive used is stored separately or mapped to an external library. Non-homogeneous networks are typically used when the EDA tool primarily deals with a structural netlist representation mapped to a technology library. Such are the EDA tools used in the post- synthesis stage like place and route, and the test or ATPG tools. However, it is fairly well accepted that homogeneous Boolean networks are highly suited for optimization problems that we typically encounter in a logic synthesis

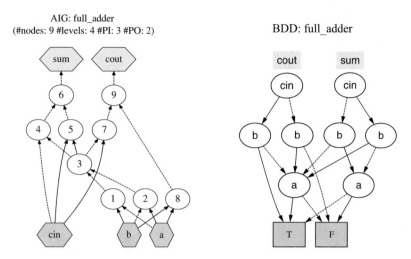

Fig. 2.1 Homogeneous Boolean networks: AIG and BDD for 1-bit full adder

tool [MCB06, MZS[+]06]. These tools greatly benefit from the regularity and simple rules[1] for easy traversal and manipulation of the underlying data structures. Note that a homogeneous Boolean network has to be a universal representation to cover the entire digital design space. The same applies to non-homogeneous Boolean network too, i.e., together with all the Boolean gates used in the network, we should be able to represent any digital design.

We use both homogeneous and non-homogeneous Boolean networks in this book. To be specific, homogeneous networks are used for synthesis problems and non-homogeneous networks are used in test. The most common homogeneous Boolean networks used in EDA tools are BDDs and AIGs. For the remainder of this book, we call the network by type name, i.e., BDD or AIG, and the non-homogeneous network as simply netlist. Figure 2.1 shows the graphical representation of BDD and AIG for a 1-bit full adder. For these networks, a negated edge is shown using a dotted line and a regular edge using a solid line.

As mentioned before, homogeneous networks have nodes formed using only one type of functionality. This functional decomposition is based on Shannon decomposition for BDDs or Boolean conjunction (a Boolean AND gate) for AIGs. A brief overview of these networks are provided next. A netlist data structure is shown in Fig. 2.2, and is covered separately in Chap. 6 on test. Note that there are several other categories of Boolean networks proposed such as Majority Inverter Graph [AGM14] and Y-Inverter Graph [CGD17b] with varying flexibility and use cases.

---

[1]These *rules* are the postulates of Boolean algebra. For more details we refer to Knuth [Knu11].

Fig. 2.2 The netlist graphical representation for a 1-bit full adder. The graph is non-homogeneous and each node has a different functionality

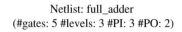

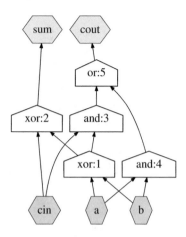

2.2.1 Binary Decision Diagrams

A BDD is a graph-based representation of a function that is based on the Shannon decomposition: $f = x_i f_{x_i} \vee \bar{x}_i f_{\bar{x}_i}$ [Bry95]. Here, the logic primitive is a multiplexer formed using Shannon decomposition. Applying this decomposition recursively allows dividing the function into many smaller sub-functions. *Reduced Ordered Binary Decision Diagram* (ROBDD) is the *canonical* version of BDDs where no sub-BDD is represented more than once. For the remainder of this book, "BDD" stands for ROBDD. BDDs are a universal representation and any Boolean function can be represented in terms of BDDs. They are unique to a given input variable order. BDDs make use of the fact that for many functions of practical interest, smaller sub-functions occur repeatedly and need to be represented only once. Combined with an efficient recursive algorithm that makes use of caching techniques and hash tables to implement elementary operations, BDDs are a powerful data structure for many practical applications. BDDs are *ordered*. This means that the Shannon decomposition is applied with respect to a given variable ordering. The ordering has an effect on the number of nodes. The number of nodes in a BDD is called its *size*. The size of the BDD varies with a different input variable order. Improving the variable ordering for BDDs is NP-complete [BW96]. However, unlike conventional applications approximate computing often relies on the input and output variable order. It is important to preserve the order especially when dealing with error metrics such as worst-case error. The worst-case error is related to the magnitude of the output vector and assigns a weight depending on the bit-order. See Sect. 2.5 for a detailed overview of error metrics used in the approximate computing. In this book, we only consider BDDs with a fixed variable ordering and assume that this order is the natural one: $x_0 < x_1 < \cdots < x_{n-1}$. The terminal nodes of a BDD are *true* and *false* represented as either 1/0 or using the notation $\top/\bot$. BDDs by itself is a vast topic. References [Bry95, HS02, DB98] and [And99] provide a detailed and self-contained introduction on BDDs. The algorithms presented in the later chapters depend on the *co-factors*, *ON-set*, and the *characteristic function* of the BDD. Therefore, these aspects are given in the subsequent sections.

2.2.1.1 Co-factors of a BDD

The co-factor of a function f wrt. to a variable x_i is the new Boolean function restricted to the sub-domain in which x_i takes the value 0 or 1. When the variable x_i is restricted to take the value 1, the resulting function is called *positive co-factor* represented as f_{x_i}. Similarly, *negative co-factor* is obtained by restricting the value of x_i to 0 and is represented as $f_{\bar{x}_i}$. In other words,

$$f_{x_i} = f_{|x_1=1} = f(x_0, x_1, \ldots, x_i = 1, \ldots x_{n-1}) \qquad (2.1)$$

$$f_{\bar{x}_i} = f_{|x_1=0} = f(x_0, x_1, \ldots, x_i = 0, \ldots x_{n-1}) \qquad (2.2)$$

Note that a co-factor is a general concept in Boolean algebra and not limited to BDDs alone. The Shannon expansion theorem[2] which forms the basis for the BDD graph-based representation is derived using co-factors.

2.2.1.2 ON-Set of a BDD

Given a Boolean function $f(x) = f(x_0, \ldots, x_{n-1})$, the ON-set refers to the number of binary vectors $x = x_0 \ldots x_{n-1}$ such that $f(x) = 1$ [Som99]. The ON-set, represented as $|f|$, gives the total number of input combinations of a BDD that evaluate to "1". Alternately, one can define an OFF-set of the BDD, which gives the total number of input vectors that evaluate to "0". The ON-set is a very useful concept in evaluating the error-rate of an approximation circuit.

2.2.1.3 Characteristic Function of a BDD

A characteristic function χ_f of a BDD is a single output function combining both its inputs and the original outputs, i.e., χ_f provides the mapping in the Boolean domain, $\chi_f : \mathbb{B}^{m+n} \to \mathbb{B}$, for a function f with n inputs and m outputs. The χ_f is defined as follows:

$$\chi_f(x_0, \ldots, x_{n-1}, y_{m-1}, \ldots, y_0) = \bigwedge_{0 \le j < m} \left(f_j(x_0, \ldots, x_{n-1}) \oplus \bar{y}_j \right), \qquad (2.3)$$

---

[2]The Shannon expansion theorem: $f = x_i f_{x_i} \vee \bar{x}_i f_{\bar{x}_i}$ is alternately called Boole's expansion theorem [Bro90].

by forcing y_j to equal the evaluation of $f_j(x_0, \ldots, x_{n-1})$. The characteristic function evaluates to true, if and only if $x_0, \ldots, x_{n-1}$ and $y_{m-1}, \ldots, y_0$ are a valid input/output pattern for f. Several BDD-based algorithms to be introduced in the subsequent chapters to evaluate the error metrics such as average-case error and worst-case error are based on the characteristic function.

2.2.2 And-Inverter Graphs

An *And-Inverter Graph* (AIG) is a homogeneous Boolean network where each node is an AND gate with two inputs and one output [MCB06]. The nodes in an AIG are formed using the Boolean conjunction operator ($\wedge$) and optional inversions at the inputs/output. An AIG is a universal representation. Similar to BDDs, any Boolean function can be represented using AIGs. However, in contrast to BDDs, AIGs are fundamentally non-canonical, i.e., they are not unique wrt. any input variable order. Canonicity of a data structure is an important property useful in satisfiability problems. AIGs can be used efficiently for logic synthesis, technology mapping, and formal verification, and are highly scalable. The terminologies *path*, *depth*, and *cut* are important in the context of an AIG. These are introduced next.

A *path* in an AIG is a set of nodes starting from a PI or a constant, and ending at a PO. The *length* of the path is the number of nodes in the path excluding the PI and PO. The *depth* of an AIG is the maximum length among all the paths and the *size* is the total number of nodes in the AIG. The depth of an AIG corresponds to delay and size corresponds to area. It must be noted that the area and the delay of the final implemented circuit heavily depends on the technology mapping, routing delay, and other technological parameters and as such cannot be deduced from synthesis itself. However, in the non-mapped network, the depth and size of AIG graph correlates well with the delay and area of the circuit.

A *cut* of an AIG node v is a set of nodes C, called *leaves*, such that (1) every path from v to a primary input must visit at least one node in C, and (2) every node in C must be included in at least one of these paths [CD96]. The node v is called the *root* of the cut and there may be several cuts for one node. The notation $|C|$ is used to denote the number of leaves. A cut is called *k-feasible*, if $|C| \leq k$. The *cut size* is the number of nodes in the transitive fan-in cone of v up to the leaves without including them. The size of the cut is also called the *area* of the cut. The *depth* of the cut is the maximum length among all the paths from the root node to any of the leaf node. Each cut expresses a *cut function* at the root expressed in terms of its leaves as inputs. Intuitively a k-feasible cut represents a single output function g, with k inputs. This function is also called the *local function* of the cut. A cut is an important concept when discussing the synthesis of approximate computing provided in Chap. 5. A cut-based algorithm called *approximation-aware rewriting* is employed to achieve automated synthesis with approximations.

AIGs have found important applications in formal verification. This is due to several reasons. AIGs are easy to manipulate and highly scalable. Coupled with simulation and Boolean satisfiability, AIGs can be used to compute functional symmetries, don't cares, and many more useful properties that aid in synthesis and formal verification [BB04, MZS⁺06, CMB05]. Besides, AIGs can be efficiently transformed to *Conjunctive Normal Forms* (CNFs) that are traditionally used in a satisfiability solver. Moreover, today, state-of-the-art SAT solvers can find solutions to problems with millions of circuit elements. The scope and reach of Boolean satisfiability and SAT solvers are constantly being expanded with new industry level applications. An overview of the Boolean satisfiability is provided next along with the other related topics necessary in the context of approximate computing.

2.3 Boolean Satisfiability

The Boolean Satisfiability (SAT) problem determines if there exists an *interpretation* that satisfies a given *Boolean formula*.[3] To elaborate, given a Boolean formula $f(x_0, \ldots, x_{n-1})$, satisfiability asks the question: does there exist any value assignment to the input variables x_0 to x_{n-1} such that $f(x_0, x_i, \ldots, x_{n-1}) = 1$. If there exists such a value assignment the formula is satisfiable; else it is unsatisfiable. Satisfiability is one of the most fundamental problems in computer science. The so-called *SAT* solvers are the devises or tools used to determine the satisfiability of a Boolean formula. Note that the SAT solver needs to find *only one* such value assignment to the variables. Modern SAT solvers can be very well taken as black-boxes with a clearly defined interface to pass on the Boolean formula and get back the results.

SAT is classically used for equivalence checking and property checking problems in the context of digital circuit verification. Equivalence checking is the problem of determining whether two different implementations of the same specification are equivalent. This is exhaustively verified for all possible input combinations providing a formal guarantee. On the other hand, property checking verifies that whether certain properties such as *liveness* hold throughout the working of the hardware. Property checking is also called model checking. This can be used, e.g., to find dead-locks in state machines. In this book, we concentrate on the equivalence checking problems using SAT. A typical use case is to encode the problem into CNF using *Tseitin transformation* [Tse68]. Further, the SAT solver traverses this CNF form and reports if the encoded problem is satisfiable or not. The SAT solver reports the input formula to be SAT, UNSAT, or ABORT. "SAT" means the SAT solver is able to find *at least* one satisfying input assignment to the formula. If

---

[3]Note: Strictly speaking a Boolean formula is different from a Boolean function. In particular, several Boolean formulas can represent the same Boolean function. For example, $x_1 \lor x_2$ and $(x_2 \land !x_1) \lor x_1$ are two distinct Boolean formulas, but evaluate to the same Boolean function. In this book, this distinction is not always respected. The terms Boolean formula and Boolean function are used interchangeably unless a clear differentiation is needed. Refer [Knu11, HS02] for a detailed treatment of Boolean algebra and related topics.

there are no such satisfying assignments, the solver returns an UNSAT. Further, an ABORT is returned when the solver is not able to determine the problem due to resource limits.

The DPLL algorithm (Davis Putnam Logemann Loveland algorithm) is classically used for CNF satisfiability problems. This algorithm uses a systematic backtracking search procedure to traverse the variable assignment space looking for satisfiable assignments. Note that the complete variable assignment space grows exponentially with the number of variables. An initial polarity on the literals (i.e., an assignment that a given variable is 1 or 0 corresponding to *true/false*) is assumed to prune the search space. Further, the solver simplifies the formula and recursively checks if the resulting formula is satisfiable or not. The solver also learns from failures on finding the solution in this search space and backtracks when needed. Recent improvements on the DPLL algorithm include *conflict driven clause learning* and branching heuristics such as VSIDS.[4] An overview of further advances made in this field can be found in [BHvMW09, vHLP07].

The significance of the problem formulation step used in the several applications of satisfiability can hardly be over emphasized. In general, SAT solvers provide nothing but a search and reasoning platform. For many applications, interpreting the given problem in terms of satisfiability is the key to finding efficient solutions. For example, *Bounded Model Checking* (BMC) is an efficient way of problem formulation to deal with the time dependent behavior of a state machine in hardware. More details on BMC are provided later in Sect. 2.3.4. The mapping or translating of the requirements to a satisfiability framework is application dependent. This book also leverages on a novel problem formulation technique called *approximation miter* to reason about the effect of approximations in a circuit. Chapters 3 and 4 explain these steps in detail.

SAT is NP-complete [Coo71]. Even then, the heuristics developed and the technical breakthroughs especially in the recent years have improved the speed and capacity of SAT solvers remarkably. As mentioned before, modern SAT solvers are employed as a black-box in applications. The improvements in the internal SAT engine is decoupled from the application. Thus, a SAT solver is essentially an *oracle* that solves a CNF problem. Current *state-of-the-art* SAT solvers can solve millions of clauses in moderate time [Sat16]. For hardware verification purposes, the underlying data structure and the techniques to convert them to the CNF are very important. The AIG data structure fits quite well in this context. There are approaches to efficiently convert the AIG to CNF form [BHvMW09]. Note that the first data structure introduced in this chapter, the BDDs are fundamentally canonical. Hence, BDDs typically do not require any CNF encoding or a SAT solver to prove the satisfiability of a formula. However, in practice, there is a considerable overhead for several circuits in order to construct the canonical BDD. This is due to the size of

---

[4]VSIDS stands for Variable State Independent Decaying Sum. Refer to [MMZ+01] for the seminal paper. VSIDS and other advances made in the field of Boolean satisfiability can be found in [BHvMW09, vHLP07].

the intermediate or final BDD and variable ordering/re-ordering scheme employed. Therefore for several practical circuits such as multipliers, equivalence checking based on AIGs is typically faster [MZS+06]. An in-depth treatment on Boolean satisfiability and associated problems can be found in [BHvMW09] and [Knu16]. Next an overview of the CNF representation and other sub-problems of the classical satisfiability problem related to approximate computing are provided.

2.3.1 CNF and Tseitin Encoding

A *Conjunctive Normal Form* (CNF) is a Boolean formula expressed as ANDs (conjunctions, the $\wedge$ operator) of ORs (disjunction, the $\vee$ operator). An example for a CNF formula is $f = (x_0 \vee x_1) \wedge (x_0 \vee x_2 \vee x_3) \wedge !x_2$. A *literal* is a variable or a negation of it. A *clause* in the CNF formula is a *disjunction* of literals. Note that the formula itself is a *conjunction* of such clauses. Thus, the CNF formula can be viewed as an AND of clauses where every clause is an OR of literals.

Tseitin encoding is traditionally used to convert a Boolean network such as an AIG to CNF form. An important advantage is that the length of the resulting formula is linear to the size of the network. For each node in the network, a new variable representing its output is introduced. Further, the expression that relates the input to the output is appended to this in the CNF form. This is repeated for all the nodes in the network, and the final network output is also encoded into a single literal and appended. As a result of Tseitin encoding the resulting CNF expression can contain more variables than the input. However, the output is equi-satisfiable[5] to the input. As an example for Tseitin encoding, an AND gate, $y = x_1 \wedge x_2$, is encoded as $(!x_1 \vee !x_2 \vee y) \wedge (x_1 \vee !y) \wedge (x_2 \vee !y)$.

Two sub-problems of the classical SAT problem which are important in the context of approximate computing are *Lexicographic SAT* (LEXSAT) and *model counting*. An overview of these topics is provided next.

2.3.2 Lexicographic SAT

A *lexicographic SAT* problem (LEXSAT) is a decision problem that returns a minimum (maximum) value of the output vector when the problem is satisfiable, or returns unsatisfiable [PMS+16]. LEXSAT finds an integer value of the output bits for a given variable ordering among all the satisfying assignments. If no such assignment exists, LEXSAT returns UNSAT similar to classical SAT. LEXSAT canonizes the classical SAT results, since the solution is unique for a given variable order. LEXSAT is NP-hard and complete for the class FP^{NP} [Kre88]. This

---

[5]Equi-satisfiable means the satisfiability properties of the input and output are the same.

complexity can be interpreted as follows: Given an oracle that solves the Boolean formula in $NP - complete$, the complexity of the LEXSAT problem is linear to that oracle. In practice this "oracle" is the classical SAT solver. In approximate computing, LEXSAT is used in the formal techniques targeted for approximations with a specification in error magnitude. Further details on the LEXSAT problem can be found in Chap. 3, where verification algorithms are proposed based on LEXSAT (see Sects. 3.3, 3.4 etc.).

2.3.3 Model Counting

In the context of Boolean satisfiability, a *model* is a distinct assignment to the input variables that evaluates the Boolean formula to be *true* (i.e., a logic "1"). The model is typically the counter-example if the formula is SAT. The *model counting* problem counts the total number of all such models. Thus, the model counting returns the number of all satisfying assignments to a given Boolean formula. However, model counting is a $\#P - complete$ problem [BHvMW09]. Hence, the computational complexity is higher compared to classical SAT. Note that this problem is alternately called *#SAT* problem. As with all the previous techniques, model counting is also dependent on the underlying data structure used. The SAT solver is commonly used for AIG-based representations. In case of a BDD, the number of satisfying assignments can be determined from the ON-set (c.f. Sect. 2.2.1.2) instead of using a SAT solver. Model counting is a very useful concept for determining certain classes of error metrics used in approximate computing such as error-rate.

As mentioned before, model counting is a $\#P - complete$ problem. Hence, for many practical scenarios the model counting tool reaches the upper limit on the run time without concluding. One alternative to this is the *statistical model counting*. As opposed to *exact* model counting, statistical approaches provide an estimate on the number of solutions. Current state-of-the-art probabilistic model counting algorithms can provide strong statistical guarantees on the estimates. Statistical model counting gives an approximate count for a given *tolerance* (ϵ) and *failure probability* (δ). The algorithm from [CMV16] is used in this work for determining the statistical error-rate of an approximation circuit (see Chap. 3 for details on error-rate computation). This algorithm relies on solution enumeration, but the number of calls to the SAT oracle is considerably reduced to logarithmic complexity using a hashing scheme. If the statistical model counting estimate is $\mathbb{E}$ and the exact result is $\mathbb{K}$, the algorithm in [CMV16] can provide a probabilistic guaranty that follow the relation:

$$Pr\left[\left(\frac{\mathbb{K}}{(1+\epsilon)}\right) \leq \mathbb{E} \leq \left((1+\epsilon)\,(\mathbb{K})\right)\right] \geq (1-\delta) \qquad (2.4)$$

See [CMV16, BHvMW09] for further details on this topic.

2.3.4   Bounded Model Checking

Bounded Model Checking (BMC) is used to check whether a circuit satisfies temporal properties or not [BCCZ99]. The temporal properties of these state machines, in principle, should be evaluated and verified for an infinite (unbounded) number of time steps. This is clearly impractical. The core idea of BMC is to place a bound on the length of possible error traces (over time) used for verification. Thus, the circuit is *unrolled* over a finite number of time steps and the resulting Boolean formula is verified using a classical SAT solver. Note that this can be done incrementally over a finite number of time steps until the computing resources are exhausted.

Property Directed Reachability (PDR) techniques can be seen as the next step in the formal verification of timed automata (state machines). PDR does the formal verification by constructing lemmas inductively, i.e., the lemmas are generated incrementally wrt. previous lemmas and dependent on each other. See [Bra13, EMB11, BHvMW09] for detailed understanding on these topics and related concepts. BMC and PDR are extensively used for verifying properties and formal equivalence of sequential approximation systems. Chapter 4 is dedicated to the formal verification of such systems.

Next an important topic on test and ATPG is reviewed. The background material presented in the next section is important in the context of test for approximate computing provided in Chap. 6.

2.4   Post-production Test, Faults, and ATPG

The manufacturing process of a circuit is vulnerable to a large number of physical *defects*, especially due to the shrinking feature sizes and process variations. These defects are directly related to the *yield* of the manufacturing process, which is the proportion of the semiconductor ICs that perform correctly. A post-production test is applied to the manufactured ICs to detect these defects and filter out the non-correct circuits. A *fault* is a logical manifestation of these defects. For example, a defect in the circuit might be an ill connected output wire to the power supply, and the corresponding logical effect (fault) is that the output value is always "1" irrespective of the inputs. *Fault models* serve as a higher level abstraction whereby logical effects of a defect can be evaluated. The most popular fault model used in practice is the *Stuck-At Fault Model* (SAFM). In this scheme, a signal connection s in the circuit is considered to be permanently "stuck" at a constant value, either 1 or 0. Several categories of defects can be detected using SAFM.

A test set T is a set of test vectors $t_1, \ldots, t_n$ applied at the circuit inputs which activates the fault location and produces a detectable difference at an observation point. In the post-production test, each detectable fault in the circuit has to be covered by at least one test pattern in the test set. The computation of this test set is called *Automated Test Pattern Generation* (ATPG) [Rot66, FS83, Lar92]. The ATPG takes all faults of a fault model as input and generates a favorably small test set with a high fault coverage.

It is well known that ATPG is a hard problem (NP-complete) and increases exponentially to the number of faults to be detected [IS75]. Hence, the *fault coverage* is very often less than 100%. Furthermore, due to the complexity of the problem, only *single stuck-at faults* are considered for pattern generation, whereby only one fault (as opposed to multiple concurrent faults) is targeted for pattern generation. Since today's circuits may contain a very large number of faults, techniques are used to reduce the number of faults which are to be targeted by the ATPG. Structural relationships are leveraged to remove dominated faults from the target fault list. ATPG can also be viewed as a Boolean satisfiability problem. Given a correct circuit and a faulty version of it, SAT solvers can be used to check if these are formally equivalent. If the result is SAT, solvers come up with a counter-example, i.e., an assignment to the inputs which invokes the fault and makes it observable at the output. This counter-example or the satisfying model can be further leveraged to generate the ATPG patterns [SE12]. On the other hand if the result is UNSAT, then it is a proof that the particular fault has no effect on circuit functioning. This approach and the associated techniques will be revisited in Chap. 6, where satisfiability is used to direct the ATPG generation for the purpose of enhancing yield of approximate computing circuits. ATPG techniques are well developed and are able to produce small test sets in reasonable time using structural implication techniques or formal proof engines [Nav11, BA02, FS83].

The *Fault Coverage* (FC) of the test set T is defined as the share of detected faults in the number of total faults. Basically, a fault can be classified by the ATPG in three categories. A fault is called *detectable*, when the ATPG proves that the fault is testable by producing a test which detects it. A fault is *redundant*, when the ATPG proves that there is no test which is able to detect the fault. A fault is classified as *aborted*, when the ATPG cannot classify the fault due to reasons of complexity. We cannot generate the test for redundant fault since these do not affect the working of the circuit. The idea of redundancy is important in the context of yield of approximate computing circuits. Enhancing the production yield by identifying redundant faults specific to an approximate computing application is the main theme of Chap. 6.

Next the important error metrics used in the context of approximate computing are explained. As mentioned before, the applications specify the error tolerance in terms of these metrics.

2.5 Quantifying Approximations: Error Metrics

Approximate Computing hinges on cleverly using controlled *inaccuracies* (in other words *errors*) in the operation for performance improvement. Hence, in approximate computing it is very important to estimate and verify the *quality* of approximations and their effects. Approximations are introduced for one main goal, i.e., to enhance performance such as to speed-up the operation, reduce area and/or power, or improve any other design metric that is of interest. Therefore, it is natural to introduce as many approximations as possible in the design, only limited by the requirements of the end application. These requirements are typically stated in terms of metrics. For example, *The result from a 32-bit Approximate Arithmetic Logic Unit (ApproxALU) should not differ from the exact one by a magnitude of "10"*, or *the branch predictions from a memory controller should be 80% accurate*, etc. Hence, the quality of approximations is based on two interrelated factors, *viz.*, the errors manifested in the final application and the performance gain. These errors are quantified in terms of error metrics.

This section covers the general error metrics used in approximate computing. There are several categories of error metrics. Since there is a diverse set of applications where approximate computing is employed, each benefits from one particular error metric or a combination of them. An approximate adder used in image processing could have an error criteria on the magnitude of errors. As long as *the total accumulated error magnitude* is within a limit, the approximate adder is good to go. However, when we use the same adder in memory controllers, the emphasis is on the *number of errors* that the adder is making, since any error irrespective of the magnitude results in a wrong memory location to be manipulated. These two application scenarios use two distinct categories of error metrics. A detailed overview on such different types of error metrics is given next. The computation of these error metrics typically follows the formal definitions provided. Hence, this will lay the foundation of the algorithms and the CAD techniques that are discussed later. Note that the second factor on the approximation quality, the performance gain due to approximation, is typically stated in terms of speed/area/power, etc. This will be explained in detail as and when required in the subsequent chapters.

Let $f : \mathbb{B}^n \rightarrow \mathbb{B}^m$ be a Boolean function with n primary inputs and m primary outputs defined in the Boolean domain $\mathbb{B}$ and $\hat{f} : \mathbb{B}^n \rightarrow \mathbb{B}^m$ be an approximation of f. Hence, there are 2^n input assignments possible for f.

2.5.1 Error-Rate

The *error-rate* is the fraction of input assignments that lead to a different output value out of total number of input assignments:

$$e_{\text{er}}(f, \hat{f}) = \frac{\sum\limits_{x \in \mathbb{B}^n} \left[f(x) \neq \hat{f}(x) \right]}{2^n} \tag{2.5}$$

The error-rate involves counting of solutions and the magnitude of the error itself is immaterial. It relates to the number of errors introduced as a result of approximation.

2.5.2   Worst-Case Error

The *worst-case error* denotes the absolute maximum deviation in the output that the approximated function can have from the original one. In this context the output is considered to have a numerical value. Hence, the output bits have a relative order of significance from the Most Significant Bit (MSB) to the Least Signification Bit (LSB). Integer values are assigned to the outputs of f and $\hat{f}$ and the error is the difference between the respective outputs. The worst-case error is related to the magnitude of the error due to approximation. This metric is also referred to as *error-significance* in literature [Bre04]. The worst-case error is given by the formula

$$e_{\text{wc}}(f, \hat{f}) = \max_{x \in \mathbb{B}^n} \left| \text{int}(f(x)) - \text{int}(\hat{f}(x)) \right|, \tag{2.6}$$

where "int" denotes the integer representation of the output bit vector.

2.5.3   Average-Case Error

The *average-case error* is defined as

$$e_{\text{ac}}(f, \hat{f}) = \frac{\sum\limits_{x \in \mathbb{B}^n} |\text{int}(f(x)) - \text{int}(\hat{f}(x))|}{2^n} \tag{2.7}$$

The average-case error denotes the average magnitude of the error introduced due to approximation, per input combination. The quantity in the numerator of Eq. (2.7) is the total arithmetic error for the approximation circuit.

2.5.4   Bit-Flip Error

The maximum *bit-flip error*, e_{bf}, is defined as the maximum hamming distance between f and $\hat{f}$.

$$e_{\text{bf}}(f, \hat{f}) = \max_{x \in \mathbb{B}^n} \left(\sum_{i=0}^{m-1} f_i(x) \oplus \hat{f}_i(x) \right), \tag{2.8}$$

where m is the width of the output bit vector. Similar to error-rate, the bit-flip error does not relate to the magnitude of the approximation error.

It is important to note that, in general, each of these error metrics are independent quantities on their own. For instance, a very high worst-case error in a design does not imply that the bit-flip error or the error-rate is high. As an example, an error at the MSB bit of an arithmetic circuit such as an adder results in a high worst-case error whereas the corresponding bit-flip error is only 1.

We conclude the discussion on the preliminaries here. We start with the automation techniques to compute the error metrics for approximation circuits in the next chapter.

Chapter 3
Error Metric Computation
for Approximate Combinational Circuits

In this chapter, the different algorithms to symbolically compute the error metrics
of approximate combinational circuits are explained. This essentially forms the
basis for the formal verification of such circuits. The error metrics introduced in
the previous chapter (see Sect. 2.5) are used to quantify the approximations in the
system. In the past, techniques based on statistical analysis have been proposed
for the error characterization of the approximation circuits. However, statistical
techniques typically depend on a probabilistic error model, input vectors, and
a probabilistic algorithm, which are application dependent, varying over time,
and difficult to predict at the design stage. Further, a complete set of input
vectors that can represent all the corner cases of the circuit operation is clearly
impractical. Hence, such techniques cannot *guarantee* a reliable circuit operation
under approximations. Since by design, approximation circuits can produce errors,
it is imperative to ensure the bounds of the errors committed. Therefore, error
analysis with formal guarantees is a must to the design of an approximate computing
hardware. In this book, we dedicate two chapters, the current one and the next
chapter, to this analysis.

This chapter covers two important aspects: First the computation of the errors in
the system and the test, if these match the specification. The second aspect is the
equivalence checking under approximations, i.e., whether two implementations of
the same specification are *equivalent* subject to a *given error metric* specification.
The techniques used for both the cases are essentially the same with some minor
differences. In the case of equivalence checking with approximations, the algorithms
need to take only a decision whether the respective error metrics are within the limit
or out-of-bound. Hence, these techniques can terminate early as soon the error limit
is violated, without the need to compute the final error metrics. Note that when the
error metrics are set to "0" (disabled), equivalence checking under approximations
reduces to classical equivalence checking.

© Springer Nature Switzerland AG 2019
A. Chandrasekharan et al., *Design Automation Techniques*
for Approximation Circuits, https://doi.org/10.1007/978-3-319-98965-5_3

In this book, the formal verification of approximate computing circuits is organized into two chapters. This chapter is dedicated to the verification of combinational circuits. The next chapter generalizes some of the important techniques developed here and applies them to sequential approximation circuits. We start with an overview of the formal verification problem under approximations.

3.1 Overview

The algorithms for computing the respective error metrics for the combinational circuits are based on the underlying data structure employed. Thus, there are different approaches based on BDD and SAT techniques to compute the error metrics. SAT techniques operate on a Conjunctive Normal Form (CNF) data structure which is typically derived from an And-Inverter Graph (AIG). Approaches based on BDD directly operate on the underlying decision diagram and do not require any explicit conversion. First an overview of the BDD-based algorithms is given and then more scalable SAT versions of the same are presented. First part of the individual sections in BDD and SAT explain the *error-rate* computation. Further, three algorithms for the BDD-based methods (Algorithms 3.1, 3.2, and 3.3) and two algorithms for the SAT-based methods (Algorithms 3.4 and 3.5) are presented. These cover the worst-case error, the average-case error, and the bit-flip error metrics.

There are two main steps in all the formal verification approaches: *problem formulation* or *problem encoding* and the *verification algorithms*. This is depicted in Fig. 3.1. In the problem formulation step, a *miter* circuit is formed incorporating the approximate computing error metrics. The miter called *approximation miter* is specific to the error metric under consideration and is formed from the approximated circuit, the golden non-approximated circuit, and the specification on the error metrics. After this step, respective verification algorithms are run on the output of this miter. These algorithms are numbered 1–5 and are shown in Fig. 3.1 under the respective categories.

The approximation miter has three main categories, *XOR approximation miter, difference approximation miter*, and the *bit-flip approximation miter*—used for error-rate, worst-case error, and bit-flip error, respectively. As shown in Fig. 3.1, this problem formulation step is common to both BDD- and SAT-based approaches. The distinction comes only in the second step: the employed algorithms. Further details on individual miter circuits and the employed algorithms are provided in the respective sections to follow.

A comparison of the SAT- and BDD-based techniques and the relative merits and demerits are provided in Sect. 3.5 towards the end of this chapter. This section details the experimental results on benchmark circuits. All the results provided here are generated from the *aXc* tool. Parts of the results have published in [SGCD16, CSGD16b].

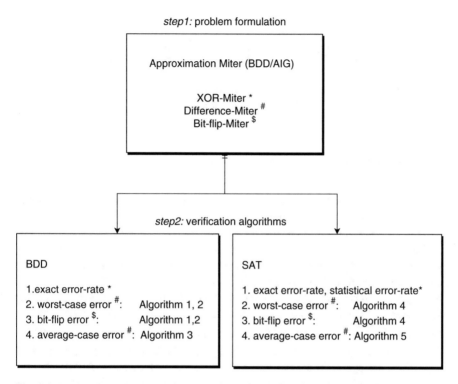

Fig. 3.1 Problem formulation and algorithms for formal verification of error metrics

3.2 BDD-Based Methods

The methodology and the relevant algorithms used for BDD-based error computation are outlined here. As mentioned before, the problem formulation and the verification algorithms are specific to the type of the error metric under consideration. Each of these are explained in the subsequent sections.

3.2.1 Error-Rate Using BDDs

For error-rate e_{er}, the approximated and the non-approximated golden circuits are represented in the BDD form and the symbolic approximation miter circuit is formed on the output bits as shown in Fig. 3.2. This miter circuit computes a bitwise exclusive OR operation on the outputs. All such output XOR bits are further OR'ed to form a single output signal y for the complete miter circuit. This miter is in many ways the same as the classical miter circuit used in conventional equivalence checking. In the context of approximate computing, this miter is termed as *XOR*

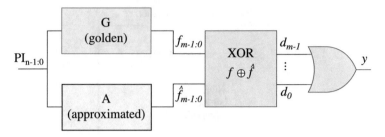

Fig. 3.2 XOR approximation miter for error-rate computation

approximation miter. This terminology differentiates the error-rate miter from other categories of approximation miters introduced in the subsequent sections.

The ON-set of the XOR-miter BDD represents all the combinations of inputs that evaluate to "1". In other words, for the miter BDD representation, the output signal will be "1" only for those input combinations where the golden and approximated circuits evaluate to a different result. Therefore, counting the ON-set of the miter BDD gives directly the total number of errors. From this total error count, the final error-rate is computed using the Definition 2.5 given in the previous chapter, i.e., $e_{\text{er}} = \dfrac{total\_errors}{2^n}$, where n is the number of input bits. Counting the ON-set forms the second step in error-rate computation. Note that standard graph traversal algorithms are available to count the ON-set of the BDD [HS02, DB98]. See the tool documentation of *aXc* for the specific implementation details used in this work.

3.2.2   Worst-Case Error and Bit-Flip Error Using BDDs

A similar procedure as in error-rate computation is used to evaluate the worst-case error and bit-flip error. The first step is to formulate the respective approximation miter circuits. In the second step, both these error metrics require a *symbolic maximum value function* that operates over BDDs. This function computes the maximum value *max* using a max algorithm and is the crux of these procedures. An overview of the steps to formulate the approximation miter in terms of BDDs is given first, followed by the details of the max algorithm.

The miter circuits that have to be formed in order to compute the worst-case error and the bit-flip error are shown in Figs. 3.3 and 3.4, respectively. These circuits are called *difference approximation miter* and *bit-flip approximation miter*. For the worst-case error, an arithmetic full subtractor and an absolute value function are formed using the output bits of the golden and the approximated BDDs. Similarly, for the bit-flip error, the individual output bits of the golden and the approximated BDDs are XOR'ed and an adder BDD function is formed which adds the individual XOR output bits. Note that the number of bits in the final output of the bit-flip approximation miter is $p = \lfloor \log_2 m \rfloor + 1$, where m is the output bit-width of the

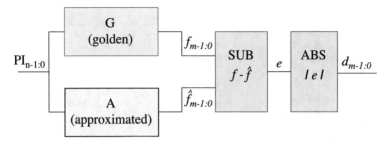

Fig. 3.3 Difference approximation miter for worst-case error computation

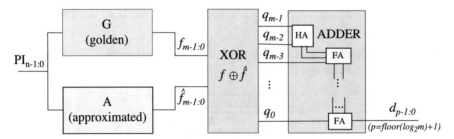

Fig. 3.4 Bit-flip approximation miter for bit-flip error computation

golden (and approximated) circuit. Thus, the respective BDDs are formed the same way as given in Eqs. (2.6) and (2.8).

The difference approximation miter symbolically represents the difference in the magnitudes due to the error introduced by approximation. Similarly, the bit-flip miter symbolically represents the number of bit-flips in the approximated circuit. Hence, the maximum value of these respective miters gives the worst-case error and the bit-flip error, respectively. The next section explains the different algorithms proposed to compute the maximum integer value possible at the output of these miter circuits.

3.2.3 Algorithms for max Function

Two algorithms named 1 and 2 are presented for the efficient computation of the maximum value max of a BDD. Algorithm 3.1 uses a bit masking technique to compute max whereas Algorithm 3.2 uses the characteristic function of the BDD to compute the same. In order to illustrate both algorithms, a Boolean function whose truth table is given in Example 3.1 is used. Note that this example function is the resulting BDD from the problem formulation explained in the previous section, i.e., the outputs of Fig. 3.3 for the worst-case error and Fig. 3.4 for the bit-flip error. For this example function, the 2-bit input is $x : \{x_1, x_0\}$ and the 5-bit output is represented as $d(x) : \{d_4, \ldots, d_0\}$. The integer interpretation of the output is k. This value is given in the last column of (3.1).

Algorithm 3.1 BDD maximum value using mask

1: **function** MAX_VALUE(BDD $d_{m-1,...,0}$)
2:     set $max \leftarrow 0$, set $i \leftarrow m$, and $\mu \leftarrow \top$
3:     **while** $i \geq 0$ **do**
4:         set $i = i - 1$
5:         set $\mu' \leftarrow \mu \wedge d_i$
6:         **if** $\mu' \neq \bot$ **then**
7:             set $max \leftarrow max + 2^i$
8:             set $\mu \leftarrow \mu'$
9:         **end if**
10:     **end while**
11:     **return** max
12: **end function**

Example 3.1 Boolean function with input $x : \{x_1 x_0\}$ and output $d : \{d_4 d_3 d_2 d_1 d_0\}$. The output value in integer is k.

$$
\begin{array}{c}
\textbf{Truth table for Example 3.1}
\end{array}
$$

input: x		output: d					int(d)
x_1	x_0	d_4	d_3	d_2	d_1	d_0	k
0	0	0	1	0	1	0	10
0	1	0	0	1	1	0	6
1	0	0	1	1	0	1	13
1	1	0	1	1	0	0	12

(3.1)

The goal of the max Algorithms 3.1 and 3.2 is to find the maximum integer output value for a given Boolean function. In other words, the algorithms should return 13 for the circuit given in Example 3.1.

3.2.3.1 Algorithm 3.1 to Find *max*

The individual steps are outlined in the listing Algorithm 3.1. The algorithm uses a Boolean function μ as mask and iterates over the columns of the truth table from the MSB function d_{m-1} to the LSB function d_0, i.e., the initial conditions are: the mask μ set to $\top$,[1] the maximum value max is set to 0, and the iterator index i is set to m (Line 2). The mask further successively iterates over the output functions from d_{m-1} to d_0 (Lines 3–4). In each iteration, a temporary mask μ' is updated with bitwise AND of the current output column d_i and the mask μ (Line 5). After this operation, the condition that μ' does not reduce to $\bot$ implies the following:

---

[1] $\top$ is terminal node ONE and $\bot$ is terminal node ZERO of the BDD.

- There exists a value "1" that contributes to max in at least one row entry for that given column d_i of the truth table.
- Only those row entries in d_i where there is a "1" can contribute further to max as we iterate the mask towards the LSB columns.

Thus, under these conditions, the mask is updated to μ' and the corresponding maximum value is added to max (Lines 6, 7, and 8). On the other hand, when the mask evaluates to $\perp$, the algorithm simply moves on to the next column since no row entry in that column can contribute to max. In this way, the input assignments that do not contribute to the maximum value are pruned out early. Algorithm 3.1 does m BDD operations altogether in m steps to compute the final max. In the last step the algorithm returns the computed max value.

To further illustrate using Example 3.1, the relevant steps for computing the maximum value after applying Algorithm 3.1 are shown in Table 3.2. The first column is the step index i initialized to 4, the MSB index of the output. The mask μ and the temporary mask μ' are shown in the second and third column in the row form. The value of max which gets updated in each iteration is given in the next column followed by a note whether temporary mask μ' is reduced to $\perp$ or not. The procedure takes 5 steps from $i = 4$ to $i = 0$, corresponding to d_4 to d_0 for computing max.

$$
\begin{array}{|c|c|c|r|c|}
\hline
\multicolumn{5}{|c|}{\text{Steps to find } max \text{ using Algorithm 3.1 in Example 3.1}} \\
\hline
\text{Step}: i & \text{Mask}: \mu & \text{Temp}: \mu' & \text{max} & \text{Comment} \\
\hline
4 & 1111 & 0000 & 0 & \mu' = \perp \\
3 & 1011 & 1011 & 2^3 = 8 & \mu' \neq \perp \\
2 & 0011 & 0011 & 8 + 2^2 = 12 & \mu' \neq \perp \\
1 & 0011 & 0000 & 12 & \mu' = \perp \\
0 & 0010 & 0010 & 12 + 2^0 = \mathbf{13} & \mu' \neq \perp \\
\hline
\end{array}
$$

(3.2)

As evident from Table 3.2, max is updated only when $\mu' \neq \perp$ in steps 3, 2, and 0. The maximum value computed is shown in bold in the final step, i.e., the row where $i = 0$. The algorithm returns this maximum value. An alternate algorithm to compute the value of max is presented in the next section.

3.2.4 Algorithm 3.2 to Find max

Algorithm 3.2 uses the characteristic function of the BDD to compute the maximum value max. The various steps are outlined in Algorithm 3.2.

The characteristic function X of the BDD is generated in the following order $X_d : \{d_{m-1}, d_{m-2}, \ldots, d_0, x_0, x_1, \ldots, x_{n-1}\}$ (Line 2). The characteristic function for Example 3.1 is depicted in Fig. 3.5. In Algorithm 3.2, the relevant node under consideration is n and the corresponding level of the node in the BDD is shown as i.

Algorithm 3.2 BDD maximum value using characteristic function

---

1: **function** MAX_VALUE_CHI(BDD $d_{m-1,...,0}$)
2:     set $\chi_{(d_{m-1},...,d_0:x_0,...,x_n)} \leftarrow characteristic(d)$
3:     set $n \leftarrow root\_node(\chi)$, set $max \leftarrow 0$ and $i \leftarrow 1$
4:     **while** $i \geq m + 1$ **do**
5:         $n_h \leftarrow high\_child(\chi_n)$
6:         **if** $n_h \neq \bot$ **then**
7:             set $n \leftarrow n_h$
8:             set $max \leftarrow max + 2^{m-i}$
9:         **else**
10:            set $n \leftarrow low\_child(\chi_n)$
11:        **end if**
12:    **end while**
13:    **return** max
14: **end function**

The starting level of the root node χ is 1. The functions $high\_child()$, $low\_child()$ are used to take the high, low children of the BDD node n.

Initially, n takes the root node of χ, max is set as 0, and the level i set to 1 (Line 3). Starting from the root node of χ, the procedure follows high edges as long as they do not lead to the zero terminal $\bot$. Whenever a high edge can be followed, the current value of max is incremented by a power of 2 that corresponds to the current level (Lines 5–8). In Fig. 3.5 this path is emphasized with thicker lines. The algorithm stops when the first node is encountered that is labeled by an input variable (Line 4). This node represents d_{max}, i.e., the function of all input assignments that evaluate to the maximum value max. The node corresponding to d_{max} is highlighted in bold in Fig. 3.5. The algorithm returns the final computed value of max. The individual steps that lead to the final max in case of the Example 3.1 are shown in thicker edges of Fig. 3.5.

The experimental results and the comparison between the max algorithms are deferred to Sect. 3.5. Next, the algorithm to compute the average-case error e_{ac} is explained. This algorithm is also based on the characteristic function χ.

3.2.5   Average-Case Error Using BDDs

The computation of the average-case error is more involved than the worst-case error.[2] The input BDD $d(x)$ for the computation of average-case error is formulated in the same way as that of the worst-case error (see Sect. 3.2 and Fig. 3.3). This BDD represents the *difference approximation miter*. In order to compute the average-case error an algorithm similar to Algorithm 3.2 with the characteristic function is used.

---

[2]Computational complexity of the respective error metrics is explained after introducing the relevant SAT-based methods in Sect. 3.3.

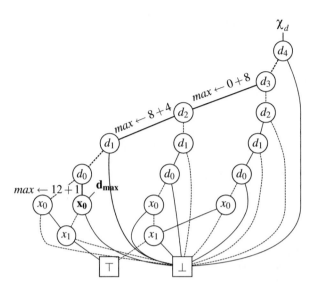

Fig. 3.5 Characteristic function of (3.1) to compute the maximum value

The average-case error is computed using the equation provided in Chap. 2, i.e.,

$$e_{\text{ac}}(f, \hat{f}) = \frac{\sum\limits_{x \in \mathbb{B}^n} |\text{int}(f(x)) - \text{int}(\hat{f}(x))|}{2^n} \tag{3.3}$$

For the purpose of computing the average-case error, we first note that the numerator of Eq. (3.3) can be written as $\sum_{x \in \mathbb{B}^n} d(x)$. Using an intermediate variable v, this can be further reformulated as

$$\sum\limits_{x \in \mathbb{B}^n} d(x) = \sum\limits_{0 \leq v < 2^m} v \cdot |\{x \in \mathbb{B}^n \mid d(x) = v\}|, \tag{3.4}$$

i.e., instead of *summing* up all function values by *iterating* over all assignments, we *iterate* over all function values and *multiply* them with the occurrence of assignments that evaluate to them. We call (3.4) a *weighted sum* and v a *path weight*. Further we give an algorithm, *viz* Algorithm 3.3 to compute the weighted sum using the BDD of χ_d.

The procedure used to compute the weighted sum is outlined in Algorithm 3.3.

As mentioned before, the input to Algorithm 3.3 is the BDD function $d(x) : \{d_{m-1}, d_{m-2}, \ldots, d_0\}$ and the result is the weighted sum $wsum$. The first step is to construct the characteristic function in the order of output variables and then the input variables, i.e., $\chi_d : \{d_{m-1}, d_{m-2}, \ldots, d_0, x_0, x_1, \ldots, x_{n-1}\}$ (Line 2). The characteristic function for Example 3.1 as applied to Algorithm 3.3 is shown in Fig. 3.6. The main concept in Algorithm 3.3 is explained next in order to facilitate easier understanding of the individual steps involved.

Algorithm 3.3 BDD weighted sum

 1: **function** WEIGHTED_SUM(BDD $d_{m-1,\dots,0}$)
 2:     set $\chi_{(d_{m-1},\dots,d_0:x_0,\dots,x_{n-1})} \leftarrow characteristic(d)$
 3:     set $wsum \leftarrow 0$ and set $n \leftarrow root\_node(\chi)$
 4:     $push(stack, n)$
 5:     **while** $not\_empty(stack)$ **do**
 6:         $n \leftarrow pop(stack)$
 7:         **if** $level(n) \geq m + 1$ **then**
 8:             set $v \leftarrow get\_path\_weight(n)$
 9:             set $wsum \leftarrow wsum + (v * ON\text{-}SET(n))$
10:         **else**
11:             $annotate\_path\_weight(n)$
12:             set $n_l \leftarrow low\_child(\chi_n)$
13:             set $n_h \leftarrow high\_child(\chi_n)$
14:             **if** $n_h \neq \bot$ **then**
15:                 $push(stack, n_h)$
16:             **end if**
17:             **if** $n_l \neq \bot$ **then**
18:                 $push(stack, n_l)$
19:             **end if**
20:         **end if**
21:     **end while**
22:     **return** $wsum$
23: **end function**

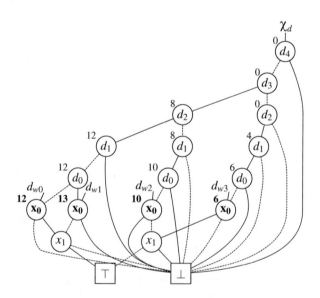

Fig. 3.6 Characteristic function of (3.1) to compute the weighted sum

In the previously explained Algorithm 3.2, it is only required to take one path in the characteristic function which leads to the maximum value and stop at the first input variable x_0 to compute max and d_{max}. However, in Algorithm 3.3, all the paths from the root node down to the level of the input variable node x_0 need to be evaluated and the path weight on all such paths has to be computed. The final path weight corresponding to the level x_0 is shown in bold in Fig. 3.6 (numbers 12, 13, 10, and 6). Further the ON-set of all the functions at this level gives the number of paths from this level that evaluates to 1. In the context of the average-case error, the ON-set of these functions (i.e., those which are marked as $d_{w0}, d_{w1}, d_{w2}, d_{w3}$ in Fig. 3.6) represents the number of ways a given error can happen. The error corresponding to this ON-set is the path weight. Therefore, the sum of products of individual path weights and the ON-set gives the weighted sum of the BDD $d(x)$. As mentioned before this weighted sum $wsum$ corresponds to the numerator in Eq. (2.7) and from this value the average-case error can be computed.

In order to efficiently compute the weighted sum, a depth first search on the characteristic function χ_d with the help of a stack data structure is used. In Algorithm 3.3, the functions $push()$ and $pop()$ pushes/pops a node to/from the stack. Further the procedure $annotate\_path\_weight()$ is used to evaluate and annotate the corresponding node with the path weight and $get\_path\_weight()$ to retrieve the already annotated path weight from the node. The path weight is computed the same way max is evaluated in Algorithm 3.2. Note that the implementation is free to use any depth first search technique and path weight evaluation method. Algorithm 3.3 is only one such approach. The functions $high\_child()$ and $low\_child()$ are used to retrieve the True and False children of the BDD node, respectively. The $ON\text{-}SET$ function returns the ON-set count of the given BDD node.

With these functions it is relatively straightforward to explain Algorithm 3.3. Initially, the stack is populated with the root node of the characteristic function χ and the $wsum$ is set to 0 (Lines 2–4). Further the top element of the stack is popped out and the level of this node is checked. If the level crosses m, i.e., the maximum level of the output variables, $wsum$ is updated with the path weight and the ON-set of the corresponding BDD (Lines 6–9). On the other hand, if the level is still within the maximum level of the output variables, the corresponding path weight is annotated, and the children of the current node are pushed into the stack (Lines 11–20). The algorithm iterates until the stack is empty (checked using $not\_empty()$ in Line 5) and the final $wsum$ is returned as the result.

The following worked out example gives further insight to the weighted sum algorithm and thereby the average-case error computation. Algorithm 3.3 applied to Example 3.1 is illustrated in Fig. 3.6. Each node starting from the root node until x_0 (shown in bold) is annotated with the corresponding path weight. The final path weights at the functions $d_{w0}, d_{w1}, d_{w2}, d_{w3}$ are 12, 13, 10, and 6, respectively. Since the function in (3.1) is injective, the ON-set for the d_{w0} to d_{w1} is all 1. Hence, the total weighted sum $wsum = 12 + 13 + 10 + 6 = 41$. From this weighted sum the average-case error is computed as $e_{ac} = \dfrac{wsum}{2^n} = \dfrac{41}{2^2} = 10.25$.

This concludes the BDD-based techniques to compute the error metrics. The next section explains the related algorithms based on Boolean satisfiability.

3.3   SAT-Based Methods

The different SAT-based approaches to compute the error metrics are presented in this section. SAT-based techniques also formulate the problem in the same way as in BDD- based approaches. However, these techniques use And-Inverter Graph as the base data structure instead of BDDs. Further, a CNF is derived from the AIG using Tseitin transformation. The main error metrics used in this work are error-rate e_{er}, worst-case error e_{wc}, bit-flip error e_{bf}, and average-case error e_{ac}. The computation of each of these error metrics is explained separately in the respective sections to follow. A brief overview of the problem formulation which is essential to the understanding of these techniques is provided in the beginning of each section followed by a detailed explanation of the algorithm.

3.3.1   Error-Rate Using SAT

The same XOR approximation miter circuit as in the BDD approach (c.f. Sect. 3.2.1 and Fig. 3.2) is used in order to compute the error-rate of the approximated circuit. The golden and approximated circuits are read in and a single output XOR signal is generated. Further the SAT-based *model counting* approaches are used to count the number of solutions for this miter. Recall that the number of solutions, i.e., the total number of ways the error miter output becomes one, gives the count of input assignments where the result from the golden circuit differs from the approximated one. Thus, counting the solutions of the miter gives the total number of errors introduced by the approximated circuit which is the numerator of Eq. (2.5). From this the error-rate can be computed.

As mentioned in the chapter on preliminaries, model counting is a well-known #*P-complete* problem (see Sect. 2.3.3 in Chap. 2). Hence, the computational complexity is high.[3] Therefore, two approaches for model counting are investigated in this work *viz* exact model counting and probabilistic model counting. The exact model counting directly follows the implementation used in [Thu06] utilizing improved problem encoding schemes and binary clause propagation. When concluded, this scheme provides the exact error count. However, the algorithm

---

[3]Note: It may appear that using BDD-based error count has lesser complexity and an upper hand compared to SAT techniques. But in this case, the whole problem formulation, i.e., creating the BDD circuits and miter, has a high computational complexity which offsets the later steps in the algorithm. A detailed study on different approaches can be found in Sect. 3.5 on experimental results.

is computationally intensive. On the contrary, probabilistic model counting gives probabilistic estimates on the number of solutions and is much faster. Probabilistic model counting algorithms take the *quality of the estimate* and the *correctness confidence* associated with the estimate as inputs along with the CNF representation of the XOR approximation miter. In this book work, the approach provided in [CMV16] is implemented for evaluating the error-rate in approximate computing. Chakraborty et al. [CMV16] provides a logarithmic statistical model counting algorithm with *Strongly Probably Approximately Correct* (SPAC) guarantees. The algorithm can provide a statistical model count between the ranges $\frac{\#SAT}{1 + \epsilon}$ and $\#SAT$ $(1 + \epsilon)$ with a correctness confidence probability. The numerator $\#SAT$ denotes the number of solutions and ϵ is the tolerance related to the quality of the estimate. Both the correctness confidence probability and the quality of the estimate are user configurable. Further details on statistical model counting can be found [CMV16]. Note that for many practical scenarios an estimate on the errors with probabilistic guarantees is a fair deal. At the minimum, a qualitative information on the committed errors can be obtained from such techniques. A comparison of run times over different benchmarks is explained in a later section on experimental results (Sect. 3.5). In the next section, the details of the worst-case error computation using SAT are provided.

3.3.2 Worst-Case Error Using SAT

The first part for the computation of the worst-case error, i.e., the problem formulation using the difference approximation miter, is the same as in the BDD counterpart shown in Fig. 3.3. A subtractor circuit followed by an absolute value function is created using the output bits of the golden and the approximated AIGs. We need to find the largest integer assignment at the output among all the input assignments for this absolute value function. This largest integer assignment is the maximum value, and in this context the worst-case error. Finding a unique maximum/minimum assignment to a set of variables in a Boolean formula is a Lexicographic SAT (LEXSAT) problem. The procedure used in this work is outlined in Algorithm 3.4.

The most important part of Algorithm 3.4 is the function $sat\_solve\_incr()$. This procedure incorporates incremental SAT solving and incremental conflict learning techniques to efficiently compute the lexicographically largest output integer assignment of a given Boolean function. As mentioned before, this Boolean function is the output of the difference approximation miter circuit shown in Fig. 3.3. The details on the individual steps in Algorithm 3.4 are explained next.

The initial conditions of Algorithm 3.4 are the maximum value max set to 0 and an output index iterator variable i assigned to $m - 1$ (Line 2). Recall that m is the number of output bits in the Boolean function $d(x)$. Further, a *SAT assumption*: 1 (*true*) is assigned to the ith output function d_i(Line 4). In the first iteration of the

---

Algorithm 3.4 SAT maximum value

 1: **function** MAX_VALUE(AIG $d_{m-1,...,0}$)
 2:     set $max \leftarrow 0$ and $i \leftarrow m - 1$
 3:     **while** $i \geq 0$ **do**
 4:         set $d_i \leftarrow 1$
 5:         set $p \leftarrow sat\_solve\_incr(d_{m-1}, \ldots, d_0)$
 6:         **if** $p = UNSAT$ **then**
 7:             set $d_i \leftarrow 0$
 8:         **else**
 9:             $max \leftarrow max + 2^i$
10:         **end if**
11:         set $i \leftarrow i - 1$
12:     **end while**
13:     **return** max
14: **end function**

---

loop, this function is d_{m-1}, the one corresponding to the MSB bit. The procedure $sat\_solve\_incr()$ solves the function d with this assumption (Line 5). Here, if the assumption is invalid the procedure returns UNSAT (Line 6). A value UNSAT implies that there is no input assignment that can drive this output bit to 1. In this case, the assumption is reverted by assigning the opposite polarity, i.e., 0 ($false$) to the corresponding output bit (Line 7). Alternately if the procedure $sat\_solve\_incr()$ returns SAT, the assumption that 1 is at the bit d_i is valid and max is updated with the corresponding value (Line 9). After every iteration of the loop the assumptions (either 0 or 1 corresponding to UNSAT or SAT) on the individual output bits starting from the MSB are recorded in the solver. Thus, the procedure $sat\_solve\_incr()$ incrementally solves d with all the previously recorded assumptions. This way the procedure *learns* the result from each SAT call. The index variable i is decremented and the loop proceeds to the next output bit until the LSB bit d_0 is reached (Line 11 and Line 3). The final value of max is returned as the result in the end.

The Algorithm 3.4 makes m SAT calls to compute max. The algorithm iterates from the MSB output position to the LSB position. Further the initial assumption on individual output bits is 1 for which the satisfiability is checked for. Hence, the algorithm is guaranteed to return a canonical maximum output value among all the satisfiable assignments for a given order of output bits of the Boolean function d. This maximum value is the worst-case error. The procedure $sat\_solve\_incr()$ does an incremental conflict analysis before invoking the actual SAT solver. Therefore, several direct conflicts with the existing assignments and circuit functionality can be pruned without invoking the computationally costlier solver. This way the run time is saved. Further, the initial search polarities of the corresponding variables in the SAT solver are assigned to *true*. This enables the solver to start the search in the solution space where the highest probability of the maximum value lies. This is very beneficial to the SAT outcome of the solver, since ultimately the solver has to find the maximum value. On the other hand, the UNSAT outcome does not affect the run time much. This is because the SAT solver has to traverse the entire search space irrespective of the initial polarity guidance for an UNSAT result.

A slightly different algorithm based on binary search can also be used to compute the max value. Here, the SAT solver is directly run after problem formulation and the initial result is noted. The magnitude of this initial result is taken as the lower bound of a binary search algorithm. The upper bound is fixed as the maximum value possible, i.e., 2^{m-1} for a Boolean function with m outputs. Further, the algorithm successively refines the bounds by assigning a middle-value to the outputs, and solving the resulting formula. If the solver returns UNSAT, the upper bound is modified to the middle-value. On the other hand if the solver returns SAT, the lower bound is modified. Thus, the binary search procedure enables the algorithm to converge on the upper and lower bounds, and this converging value is returned as the max. This variation of the max algorithm is further explained and elaborated later in the respective contexts in Chaps. 4 and 5 (see Algorithms 4.1 and 5.3 in the respective chapters).

3.3.3 Bit-Flip Error Using SAT

The bit-flip error computation also uses the circuit formulation similar to the BDD shown in Fig. 3.4. The same algorithm, i.e., Algorithm 3.4, and the associated techniques are used to compute the bit-flip error. In essence, only the problem formulation changes for this error metric. The bit-flip approximation miter shown in Fig. 3.4 is used instead of the difference-miter. Everything else introduced before in the case of worst-case error is applicable here too. Hence, these details are not re-iterated in this section.

The technique to evaluate the average-case error is explained in the next section. Note that the average-case error determination circuit uses the same configuration as in the worst-case error whose details are already given in the previous section (i.e., Fig. 3.3). Hence, only the relevant algorithm is explained in the subsequent section and this circuit formulation is omitted.

3.3.4 Average-Case Error Using SAT

The average-case error metric is computationally more expensive than all the previous error metrics. The major bottleneck in computing average-case error is the numerator of Eq. (2.7), the total error magnitude, which is the sum of errors over all input assignments. Recall that from Eq. (3.4) the total error magnitude can be reformulated as a weighted sum. Hence, we need to (1) find the *magnitude* of all distinct errors and (2) find the *number of occurrences* of all such errors. Further the individual error values need to be multiplied with its corresponding count of occurrence and summed up. Thus, computing the weighted sum involves both *finding* the errors (i.e., Algorithm 3.4 and other techniques) and *counting* the

---

Algorithm 3.5 SAT weighted sum

1: **function** WEIGHTED_SUM(AIG $d_{m-1,...,0}$)
2:     set $wsum \leftarrow 0$
3:     set $max \leftarrow next\_worst\_case()$
4:     **while** $max \neq 0$ **do**
5:         set $count = count\_solutions(max)$
6:         set $wsum \leftarrow wsum + (max * count)$
7:         set $max \leftarrow next\_worst\_case()$
8:     **end while**
9:     **return** $wsum$
10: **end function**

---

error occurrences (i.e., SAT-based model counting). A SAT-based algorithm listed in Algorithm 3.5 does these two main steps and returns the weighted sum as the result. From this the average-case error is computed.

In Algorithm 3.5 two procedures $next\_worst\_case()$ and $count\_solutions()$ are used to compute the weighted sum. The function $next\_worst\_case()$ applies the Algorithm 3.4 incrementally and computes the *current* worst-case error. After computing this value, the procedure further invalidates this result so that the successive calls to Algorithm 3.4 return the next worst-case error only (i.e., the next worst-case error in *magnitude*). Thus, using the function $next\_worst\_case()$ all the errors in descending order of magnitude are obtained one by one. On the other hand, $count\_solution()$ provides the number of solutions corresponding to this error magnitude. Here, the exact model counting is used without any probabilistic assumptions.[4] Note that $next\_worst\_case()$ and $count\_solution()$ operate directly on the difference approximation miter network.

The maximum worst-case error in Algorithm 3.5 is evaluated after initializing the weighted sum variable $wsum$ to 0 (Lines 2 and 3 in Algorithm 3.4). There is no error in the circuit if this value max is equal to 0 and the procedure directly returns $wsum = 0$. When there is an integer value in max, the number of occurrences of this error magnitude $count$ is computed using $count\_solution()$ (Line 5). The value $wsum$ is updated with the product of max and $count$ and the algorithm proceeds further with the next error in the descending order (Lines 6 and 7). The loop iterates until no more errors are present and the final $wsum$ is returned as the result. From the $wsum$, the average-case error is computed by applying Eq. (2.7).

---

[4]Note: It is relatively straightforward to use a probabilistic model counting approach here. However, the resulting statistical properties cannot be directly deduced since the individual probability distributions resulting from each iteration of the while loop inside Algorithm 3.5 have to be taken into consideration. This does not fall in the purview of this book work.

3.4 Algorithmic Complexity of Error Metric Computations

At this point it is relatively straightforward to establish the computational complexity of the proposed algorithms to evaluate the approximation error metrics. These results directly follow the Boolean satisfiability procedures outlined before. In fact, SAT is the first problem proven to be *NP-complete* [Coo71] and is widely used to prove the computational complexity of the other algorithms.

1. Error-rate is computed using the direct application of SAT model counting. Hence, the error-rate computational complexity is *#P-complete*. The complexity of model counting is a well-known result and the proof is available in [BHvMW09].
2. The worst-case error and the bit-flip error employ a procedure to find the canonical maximum value at the output of a Boolean function. This procedure is called the lexicographic SAT or LEXSAT. The complexity of the LEXSAT procedure is informally stated as follows:

   - *The complexity of a LEXSAT problem is linear to an oracle of complexity NP-complete.*

   The proof of the above statement can be found in [Kre88]. The *oracle* that solves the problem in NP-complete is the *SAT solver* in the maximum value procedure. Algorithm 3.4 uses m calls to the SAT solver, making it linear to the number of output bits. Hence, summing up, the complexity of the maximum value Algorithm 3.4 is FP^{NP}. Here, the notation FP (related to function problem) denotes that the LEXSAT is more involved than a plain decision problem.

   A slightly different version of the maximum value algorithm using binary search is outlined in the last paragraph of the section on Algorithm 3.4. It is straightforward to see that the number of SAT calls needed is logarithmic to the *maximum magnitude* of the output value, due to the binary search procedure involved. In both approaches the *oracle* that solves the problem is in *NP-complete*, i.e., the complexity of the SAT solver is the same.

The different techniques and various algorithms based on formal approaches to evaluate errors for approximate computing conclude here. In the next section, the experimental setup, implementation details, and the results are provided.

3.5 Implementation

All the algorithms are implemented in the *aXc* framework. The *aXc* tool reads the RTL descriptions of the golden and approximated designs in various formats including Verilog. The different error metrics are computed and reported using the command report_error. There are several options for this command (report_error -h to show help) to invoke the different techniques and individual error metrics.

3.5.1   Experimental Results

The experimental evaluation is organized into two parts. First, the different error metrics reported for several architecturally approximated ad-hoc adder designs are studied. Note that, irrespective of SAT- or BDD-based techniques, the error behavior of these circuits is the same. Besides, most of these circuits are very small and the run-time differences between the different algorithms are not significant. For evaluation purposes, we use the SAT-based approaches exclusively for this set of circuits. The run-time and scalability comparison of BDD vs SAT approaches are taken up separately on the second part of the experimental results in a bigger set of benchmarks. Here, the different BDD-based approaches are compared wrt. SAT-based approaches on standard benchmarks such as ISCAS-85 [HYH99] and EPFL circuits [AGM15]. The main comparison point of these experiments is the computational effort and scalability in terms of run time.

3.5.1.1   Error Metrics for Approximation Adders

The different error metrics for a wide range of approximate adder architectures taken from the literature are evaluated. In these adders, the approximation is introduced in the design stage itself by means of architectural approximation schemes. Several types and configurations of such adders have been proposed in the literature depending on the target application and the required error characteristics. These include *Almost Correct Adder* (ACA adder) [KK12], *Error Tolerant Adder* (ETA Adder) [ZGY09], *Gracefully Degrading Adder* (GDA Adder) [YWY+13], and *Generic Accuracy Configurable Adder* (GeAr Adder) [SAHH15]. A collection of these approximate adder architectures are available in the public repository [GA15]. Using the *report_command* in the *aXc* tool, the error characteristics of these adders are reported. As previously mentioned the evaluation is carried out only with the SAT-based techniques. Tables 3.1 and 3.2 summarize the error metrics computed for these circuits. The short-form notations used are as given in the repository [GA15]. Both 8-bit and 16-bit versions of the adders are given in the table. The number of gates and delay shown in the second and third column is taken from academic tool ABC [MCBJ08] after synthesis. This is followed by different combinational error metrics from *aXc*. Note that the error-rate given in last column of Table 3.1 is the exact error-rate. Each category has a *Ripple Carry Adder* entry given at the end. This is the normal non-approximated adder and serves as golden reference model.

As evident from Table 3.1, different approximate adder architectures demonstrate a wide range of error behavior. For example, among the several 8-bit *Gracefully Degrading Adders* [YWY+13], GDA_St_N8_M8_P6 has the lowest error-rate of 0.39%, but with a high worst-case error of 128. This may be compared with other adders of the same type, e.g., GDA_St_N8_M4_P2 (the first entry in the same category) has a lower worst-case error of 64 though with a higher error-rate of 18.75%. This design also has a much reduced delay and gate count compared to

Table 3.1 Error metrics for 8-bit approximation adders

Architecture details			Error metrics			
Circuit[‡]	Gates[†]	Delay[†]	e_{wc}	e_{ac}	e_{bf}	e_{er} (%)
8-bit adders						
Almost Correct Adder [KK12]						
ACA_II_N8_Q4	36	7.00	64	7.5	5	18.75
ACA_I_N8_Q5	52	7.00	128	3.5	4	4.69
Gracefully Degrading Adder [YWY[+]13]						
GDA_St_N8_M4_P2	35	6.80	64	7.5	5	18.75
GDA_St_N8_M4_P4	49	8.10	64	1.5	3	2.34
GDA_St_N8_M8_P1	26	4.00	168	31.5	7	60.16
GDA_St_N8_M8_P2	34	5.80	144	15.5	6	30.08
GDA_St_N8_M8_P3	46	6.20	128	7.5	5	12.50
GDA_St_N8_M8_P4	53	7.90	128	3.5	4	4.69
GDA_St_N8_M8_P5	58	9.80	128	1.5	3	1.56
GDA_St_N8_M8_P6	66	10.80	128	0.5	2	0.39
Accuracy Configurable Adder [SAHH15]						
GeAr_N8_R1_P1	26	3.80	168	31.5	7	60.16
GeAr_N8_R1_P2	35	5.40	144	15.5	6	30.08
GeAr_N8_R1_P3	46	7.00	128	7.5	5	12.50
GeAr_N8_R1_P4	52	7.00	128	3.5	4	4.69
GeAr_N8_R1_P6	44	8.60	128	0.5	2	0.39
GeAr_N8_R2_P2	36	7.00	64	7.5	5	18.75
GeAr_N8_R2_P4	40	8.60	64	1.5	3	2.34
Ripple Carry Adder						
RCA_N8	43	10.20	0	0	0	0.00

[‡]Circuits from public repository: *http://ces.itec.kit.edu/1025.php*
Naming conventions used are as given in the above repository
[†]After synthesis in ABC[MCBJ08] with command *resyn*, library mcnc.genlib
e_{wc}:worst-case, e_{ac}:average-case, e_{bf}:bit-flip errors, e_{er}:error-rate (%)
e_{wc}, e_{ac}, e_{bf}, e_{er} using aXc tool *(report_error –wc –ac –bf –er)*
Error-rate e_{er} is the exact error-rate
All error metrics are computed with SAT-based approaches

GDA_St_N8_M8_P6. In general, the designer has to evaluate all such parameters along with the error tolerance of the application before finalizing on a particular architecture.

The study of error metrics for these ad-hoc approximation adders by itself is significant. However, this is not the main focus of our work. Most of these circuits are smaller and do not provide useful insight into the computational effort required for the different SAT- and BDD-based approaches discussed before. Hence, a detailed comparative study of the proposed algorithms using other bigger benchmarks is provided in the second part of this experimental evaluation, details of which are explained next.

Table 3.2 Error metrics for 16-bit approximation adders

Architecture details			Error metrics			
Circuit[‡]	Gates[†]	Delay[†]	e_{wc}	e_{ac}	e_{bf}	e_{er} (%)
16-bit adders[‡]						
Almost Correct Adder [KK12]						
ACA_II_N16_Q4	72	7.00	17,472	2047.5	13	47.79
ACA_II_N16_Q8	93	10.20	4096	127.5	9	5.86
ACA_I_N16_Q4	102	7.00	34,944	2047.5	13	34.05
Error Tolerant Adder [ZGY09]						
ETAII_N16_Q4	72	7.00	17,472	2047.5	13	47.79
ETAII_N16_Q8	93	10.20	4096	127.5	9	5.86
Gracefully Degrading Adder [YWY+13]						
GDA_St_N16_M4_P4	99	10.00	4096	127.5	9	5.86
GDA_St_N16_M4_P8	120	10.60	4096	7.5	5	0.18
Accuracy Configurable Adder [SAHH15]						
GeAr_N16_R2_P4	84	8.60	16,640	511.5	11	11.55
GeAr_N16_R4_P4	93	10.20	4096	127.5	9	5.86
GeAr_N16_R4_P8	97	11.80	4096	7.5	5	0.18
GeAr_N16_R6_P4	94	10.20	1024	31.5	7	3.08
Ripple Carry Adder						
RCA_N16	102	13.40	0	0	0	0.00

[‡]Circuits from public repository: *http://ces.itec.kit.edu/1025.php*
Naming conventions used are as given in the above repository
[†]After synthesis in ABC[MCBJ08] with command *resyn*, library mcnc.genlib
e_{wc}:worst-case, e_{ac}:average-case, e_{bf}:bit-flip errors, e_{er}:error-rate (%)
e_{wc}, e_{ac}, e_{bf}, e_{er} using aXc tool *(report_error –wc –ac –bf –er)*
Error-rate e_{er} is the exact error-rate
All error metrics are computed with SAT based approaches

3.5.1.2 Scalability and Generality

In the second part of the experimental evaluation, we study the scalability and generality of our approaches. Two different sets of benchmark circuits are used to compare the BDD-based error metric computation and the SAT-based approaches. The first set consists of the ISCAS-85 circuits available in [HYH99]. The second set is taken from the EPFL benchmarks [AGM15]. EPFL circuits consist of very large combinational circuits. Each of these circuits is approximated by means of an automated approximation synthesis technique described in [CSGD16a]. This approximation synthesis technique is also integrated into the *aXc* tool and is the focus of Chap. 5. For now, this feature is used as it is without bothering about the underlying synthesis techniques, since the focus is on formal verification. The run time for error metric computation is limited to 2 h in the experiments. The experiments are carried out on an Octa-Core Intel Xeon CPU with 3.40 GHz and 32 GB memory running Linux 4.1.6.

Table 3.3 Summary of run-times for ISCAS-85 benchmark circuits

Benchmark details			
Circuit	Function	#PI/#PO	Gates†
c17*	Example circuit	5/2	2
c432*	27-channel interrupt controller	36/7	109
c499*	32-bit SEC circuit	41/32	163
c880*	8-bit ALU	60/26	185
c1355*	32-bit SEC circuit	41/32	155
c1908*	16-bit SEC/DED circuit	33/25	105
c2670*	12-bit ALU and controller	233/140	377
c3540*	8-bit ALU	50/22	640
c5315*	9-bit ALU	178/123	1273
c6288*	16 x 16 multiplier	32/32	1555
c7552*	32-bit adder/comparator	207/108	1175

Circuit	Run times: SAT method (s)					Run times: BDD method (s)					
	$t_{er\text{-}prb}$	$t_{er\text{-}ext}$	t_{wc}	t_{bf}	t_{ac}	t_{er}	t_{wc}	$t_{wc\text{-}chi}$	t_{bf}	$t_{bf\text{-}chi}$	t_{ac}
c17*	0.00	0.00	0.00	0.00	0.00	0.00	0.00	0.00	0.00	0.00	0.00
c432*	0.20	0.05	0.01	0.01	0.52	0.01	0.04	0.09	0.06	0.03	0.19
c499*	0.17	⋆	0.03	0.03	⋆	0.04	237.92	493.86	0.06	0.01	487.64
c1355*	0.32	⋆	0.03	0.04	⋆	0.04	310.75	575.74	0.06	0.01	588.82
c1908*	0.15	68.36	0.02	0.02	104.02	0.01	0.16	38.17	0.23	0.11	66.01
c2670*	4.59	⋆	0.01	0.01	⋆	⋆	⋆	⋆	⋆	⋆	⋆
c3540*	0.37	4762.87	0.04	0.04	326.92	1.84	19.17	326.08	10.98	22.2	392.04
c5315*	6.33	⋆	0.03	0.03	⋆	⋆	⋆	⋆	⋆	⋆	⋆

#PI, #PO: number of inputs, outputs
$\dagger$Gate count after synthesis in ABC [MCBJ08] with command *resyn*, library *mcnc.genlib*
All circuits approximated with an automated approximation synthesis tool from [CSGD16a]
Note: * in circuit names shows this approximation
$t_{er\text{-}prb}$: run time for SAT probabilistic error-rate
$t_{er\text{-}ext}$: run time for SAT exact error-rate. t_{er}: run time for BDD error-rate
t_{wc}: run time for worst-case error (SAT Algorithm 3.4, BDD Algorithm 3.1)
t_{bf}: run time for bit-flip error (SAT Algorithm 3.4, BDD Algorithm 3.1)
t_{ac}: run time for average-case error (SAT Algorithm 3.5 and BDD Algorithm 3.2)
$t_{wc\text{-}chi}$, $t_{bf\text{-}chi}$: run times for worst-case error, bit-flip error
using characteristic function (only for BDD method using Algorithm 3.2)
All run times in *seconds*. All runs set with a timeout of 2 h
'⋆' shows time-out (unsuccessful completion in 2 h run time)

A comparative study of these benchmarks is given next. The main comparison for these techniques is the run time summarized in Tables 3.3 and 3.4. Other important details such as the gate count for individual circuits are also provided alongside in the respective tables.

Table 3.4 Summary of run-times for EPFL benchmark circuits

Benchmark details			Run time: SAT (s)			Run time: BDD (s)		
Circuit	#PI/#PO	Gates†	$t_{er\text{-}prb}$	t_{wc}	t_{bf}	t_{er}	t_{wc}	t_{bf}
Arithmetic circuits								
Adder*	256/129	766	5.02	0.09	2.20	⋆	⋆	⋆
Barrel shifter*	135/128	2703	0.78	0.05	0.08	⋆	⋆	⋆
Divisor*	128/128	53,063	4.32	0.45	0.59	⋆	⋆	⋆
Hypotenuse*	256/128	15,8913	43.93	⋆	⋆	⋆	⋆	⋆
Log2*	32/32	19,327	15.58	9.47	91.23	⋆	⋆	⋆
Max*	512/130	2931	43.07	0.14	2.35	⋆	⋆	⋆
Multiplier*	128/128	16,538	⋆	⋆	⋆	⋆	⋆	⋆
Sine*	24/25	4084	1.19	0.19	0.27	2.00	14.59	33.29
Square-root*	128/64	19,408	14.36	1.56	23.15	⋆	⋆	⋆
Square*	64/128	14,351	1.49	7.10	1.47	⋆	⋆	⋆
Random/control circuits								
Round robin arbiter*	256/129	6710	8.51	0.04	0.11	⋆	⋆	⋆
Alu control unit*	7/26	112	0.00	0.00	0.00	0.00	0.00	0.00
Coding-cavlc*	10/11	499	0.01	0.00	0.00	0.00	0.00	0.00
Decoder*: 8/256	430		0.00	0.01	0.05	⋆	⋆	⋆
i2c controller*	147/142	1077	1.10	0.01	0.00	⋆	⋆	⋆
Int to float converter*	11/7	168	0.01	0.00	0.00	0.00	0.00	0.00
Memory controller*	1204/1231	34,542	338.58	1.56	697.98	⋆	⋆	⋆
Priority encoder*	128/8	795	0.86	0.00	0.00	⋆	⋆	⋆
Lookahead XY router*	60/30	234	0.16	0.00	0.00	⋆	⋆	⋆

#PI, #PO: number of inputs, outputs.   *Circuits approximated using [CSGD16a]
$\dagger$Gate count after synthesis in ABC [MCBJ08] using *resyn*, mcnc.genlib
$t_{er\text{-}prb}$: SAT probabilistic error-rate, t_{er}: BDD error-rate
t_{wc}: worst-case error (SAT Algorithm 3.4, BDD Algorithm 3.1)
t_{bf}: bit-flip error (SAT Algorithm 3.4, BDD Algorithm 3.1)
All runs with run-time limit of 2 h. ⋆ shows timeout (non-completion)

3.5.1.3   ISCAS-85 Benchmark

The results from the ISCAS-85 benchmarks are provided in Table 3.3. The first four columns are the benchmark details such as circuit name, brief functionality, inputs/outputs, and the gate count. Next, the run times for SAT-based approaches are given followed by the results from the BDD-based techniques. The time taken is shown in seconds. The columns t_{er-prb} and t_{er-ext} are tool run times for SAT-based probabilistic and exact error-rates. The columns t_{wc}, t_{bf}, and t_{ac} are the run times for worst-case error (e_{wc}) bit-flip error (e_{bf}), and average-case error (e_{ac}), respectively. For the SAT-based approaches these are computed using Algorithm 3.4 and for the BDD-based techniques Algorithm 3.1 is used. The run times t_{wc-chi} and t_{bf-chi} correspond to the worst-case error and bit-flip error computation using the BDD Algorithm 3.2.

1. Error-rate comparison: From Table 3.3, the error-rate computation using the SAT-based probabilistic technique is the fastest. The run times are several orders of magnitudes smaller compared to the SAT exact error-rate and BDD-based approaches. However, between the exact approaches, the BDD-based error-rate computation is faster than the SAT counterpart.
2. Worst-case and bit-flip error comparison: For the worst-case error and bit-flip error, SAT techniques are significantly faster than the BDD counterparts. All the benchmarks conclude in under a second for SAT while for BDDs the computations are pushed to the time-out limit in many cases.
   If we take the BDD results alone, Algorithm 3.1 is faster than Algorithm 3.2, i.e., the time taken for t_{wc} and t_{bf} is lower than its Algorithm 3.2 counterparts t_{wc-chi} and t_{bf-chi}. In Algorithm 3.2, we build the characteristic BDD where the POs get appended on top of the PIs in a specific order. Hence, the characteristic function has more levels than the input BDD and is subsequently more complex. Therefore, it is more computationally involved to build and operate on such BDDs. Such problems do not occur in Algorithm 3.1. Here, new BDDs are created but sharing the same levels and consequently the size does not increase drastically compared to Algorithm 3.2. This is the main reason why most of the benchmarks in this approach are faster.
3. Average-case error comparison: The average-case error computation has time-out in many benchmark circuits both for SAT- and BDD-based approaches. However, comparatively BDD approaches are far better than the SAT counterparts.

The main bottleneck with the BDD-based techniques is building the problem formulation circuit itself (i.e., the circuits in Figs. 3.2, 3.3, etc.). Once the circuit formulation of the individual error metrics is completed, BDD-based techniques can compute the error values relatively faster. This is true especially for error metrics that require counting in solution space such as error-rate. In comparison, SAT does not have a problem with the problem formulation circuit. Instead SAT time-outs usually happen with the algorithm employed. Hence, if we take bigger circuits, BDDs reach the limit of scalability in the problem formulation itself, rather than the respective algorithms. Note that in order to preserve the input/output bit order we do not allow dynamic variable ordering while building the BDD. In this aspect, SAT techniques scale better in terms of circuit size. This will be more evident in the second set of results with bigger EPFL benchmark circuits. These results are provided in the next section.

3.5.1.4 EPFL Benchmark

The results with the EPFL circuits are given in Table 3.4. There are two categories of circuits, arithmetic and random/control circuits. The first section of the table consists of the benchmark details followed by run times for SAT-based techniques and then the BDD-based algorithms. It can be seen from Table 3.4 that the majority of the

BDD-based error metric computations have timed out. The EPFL circuits are bigger both in gate count and the number of PI/POs. Hence, the BDD-based approaches have reached their scalability limit for such circuits. Besides, the average-case error also does not conclude in most of the bigger circuits for both SAT and BDD techniques. Hence, this is also omitted from Table 3.4.

In Table 3.4, the run times for error-rate, worst-case errors, and bit-flip errors are given for the respective SAT- and BDD-based techniques. The results of the probabilistic SAT-based error-rate computation are given in column 3 ($t_{er}(prb)$) followed by the run times for worst-case error and bit-flip error (both using Algorithm 3.4). The calculation of the exact error-rate on most of these circuits has timed out. Hence, the numbers for SAT-based exact error-rate computations are not shown separately in the table. To compute the BDD-based worst-case error and bit-flip error (the last two columns), we have used Algorithm 3.1. The fastest among all the error metrics is the SAT-based worst-case error computation (column 4). The worst-case computation takes only a few seconds for several of the EPFL circuits. Moreover, as evident from Table 3.4, the probabilistic error-rate for most of these circuits are easily computed and the run times are within the time-out limits. In general, it can be concluded that for most of these benchmarks SAT-based approaches for computing the worst-case error and bit-flip error are significantly faster than the corresponding BDD-based algorithms. Furthermore, SAT-based error-rate computation provides an option for probabilistic computation, where the computational complexity is prohibitively high to attempt an exact error-rate approach.

3.6   Concluding Remarks

In this chapter, we proposed several algorithms, based on BDD and SAT, to compute the error metrics of approximate combinational circuits. In comparison to the conventional techniques which are based on statistical methods, our approaches provide a crucial and much needed aspect of the approximation circuit design—*guarantees* on the bounds of the errors committed. The experimental results confirm the applicability of our proposed methodologies. The next chapter extends these concepts and apply them to the formal verification of sequential circuits.

Chapter 4
Formal Verification of Approximate Sequential Circuits

The second part of the formal verification problem is related to the algorithms and techniques for sequential approximated circuits. This chapter explains the methodologies for verification of such circuits. The approximation miter concept introduced in the previous chapter is generalized to include the state relationship of sequential circuits. To the best of our knowledge, very few formal verification techniques have been proposed before for the verification of approximate sequential circuits. The most closely related one to our work is the ASLAN framework [RRV$^+$14], which employs a user-specified quality evaluation circuit in the form of a test bench. However, this is a semi-automated approach, since the user has to design the quality evaluation circuit and provide as an input to the ASLAN framework. This requires a detailed understanding of the design concepts along with a sound knowledge on formal property checking (e.g., specification of a liveness property). On the contrary, our approach is fully automated. Furthermore, the case studies provided later in this chapter show that the error analysis of the approximate sequential circuits can lead to a different conclusion altogether, when compared with the analysis of combinational circuits in isolation.

In this context, it is important to differentiate an approximate sequential circuit from a non-approximated one. The difference between an approximated sequential circuit and a normal sequential circuit is that the former one has approximated combinational components as building blocks. However, the state relationship among the sequential elements is not altered.[1] Hence, the formal verification of approximated sequential circuits is essentially *proving* the *error behavior* of the system over *time*. We have developed methodologies based on Boolean satisfiability to aid this analysis. The method has been published in [CSGD16b]. Similar to the previous chapter, we start with an overview of the formal verification problem for approximate sequential circuits.

---

[1]Note that though this distinction is followed in this book, there is no universal consensus on what constitutes an approximated sequential circuit.

© Springer Nature Switzerland AG 2019
A. Chandrasekharan et al., *Design Automation Techniques for Approximation Circuits*, https://doi.org/10.1007/978-3-319-98965-5_4

4.1 Overview

The general approach for verifying the approximated sequential circuits is essentially similar to that for combinational circuits. First step is the problem encoding in the form of an approximation miter and then the verification algorithms are applied to this circuit formulation. However, there are several important differences on how an error manifest itself, making the sequential verification unique and different from that of combinational circuits.

Precisely determining the error metric of a combinational circuit using the methods provided in the previous chapter is already very helpful in the design of approximate computing, but the obtained numbers may not be accurate when considering approximated components in sequential circuits. As an example, although the worst-case error can be computed for the approximated component in isolation, the accumulated worst-case error in the sequential circuit may differ significantly, since only a subset of the input vectors may actually be assignable. Further, the sequence of successive input patterns for the approximated component depends on the sequential logic and composition of the overall circuit. In this book, algorithms based on model checking are proposed that are able to prove the absence of errors such as accumulated worst-case, maximum worst-case, or average bit-flip error—although theoretically possible when only considering the approximated component in isolation. For this purpose, the concept of an approximation miter from the previous chapter is generalized to include the state information. This sequential approximation miter incorporates both error computation and accumulation over time in sequential circuits. The state-of-the-art bounded model checking techniques are then used to prove the presence or absence of an error. In other words, the exact error metric computation problem is reduced to a decision or optimization problem using formal techniques.

When sequential elements are present in the design, the number of states for equivalence checking increases exponentially and becomes easily non- tractable. To deal with this state explosion problem several approaches have been proposed. One such approach is *Bounded Model Checking* (BMC) [BCCZ99] where the length of the sequence over which the formal verification is employed is finite. When applied to sequential equivalence checking the circuit is unrolled to a predefined number of time frames and the overall problem including the properties is solved using a SAT solver. This approach is feasible as the current state-of-the-art SAT solvers [Sat16] can solve millions of clauses in moderate time. Besides BMC, tools have evolved to use *inductive reasoning* [ZPH04] and more recently PDR [EMB11, Bra13] techniques which give vast improvements over the original method. We use BMC and PDR for the formal verification of approximated sequential circuits. These methods can precisely determine the error behavior of sequential circuits.

The details are explained as follows. At first, the general idea of an approximation miter which is adapted for sequential approximated circuits is presented. Afterwards the individual components of the approximation miter and its generic configurable form are outlined. Altogether, we can ask different questions which can be precisely answered based on suitable configurations of the approximation miter and the reduction to decision or optimization problems.

4.2 General Idea

The proposed approach introduced in this section can precisely compute accumulated errors of combinational components in sequential circuits. The main idea is to apply model checking to a *sequential approximation miter* which is illustrated in Fig. 4.1. The seq-approximation miter consists of five circuits: (1) the non-approximated original circuit C, (2) the circuit $\hat{C}$ that is obtained by replacing some combinational components with approximated implementations, (3) a circuit E to compute the error of $\hat{C}$'s outputs with respect to C's outputs and a given error metric, (4) an accumulator A that accumulates the computed errors in each cycle, and (5) a decision circuit D. Note that E and D are stateless circuits, whereas C, $\hat{C}$, and A are circuits with state information. Further, the state relationships of C and $\hat{C}$ are identical. The decision circuit D exposes the single output "bad" of the approximation miter, which is 1 if and only if the accumulated error violates a given requirement. Formal model checking techniques can be applied to show that the output signal never evaluates to 1 either unbounded or within some given number of cycles.

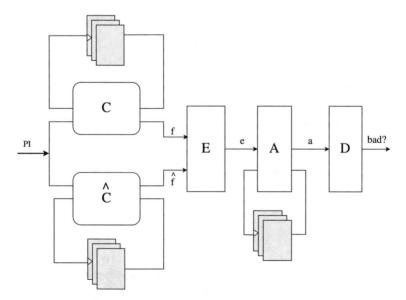

Fig. 4.1 General idea of a sequential approximation miter

4.2.1 Sequential Approximation Miter

This section illustrates the components of the approximation miter in detail. Note that these are only selected examples that can all be exchanged by user defined implementations to create custom scenarios. Some example scenarios based on the selected implementations are presented at the end of this section.

4.2.1.1 Error Computation

The stateless error computation E takes as input the output results of C and $\hat{C}$, accessible through the wires f and $\hat{f}$. Not all outputs are of interest necessarily and therefore not all outputs of C and $\hat{C}$ may be contained in f and $\hat{f}$. Further, the order of bits is set to adhere the desired numeric interpretation.

For worst-case error and average-case type computations, E is implemented as

$$e = \left| \text{int}(f) - \text{int}(\hat{f}) \right|. \tag{E1}$$

For bit-flip error computations, E is

$$e = f \oplus \hat{f} \tag{E2}$$

or

$$e = \sum_{i=1}^{m} \left(f_i \oplus \hat{f}_i \right), \tag{E3}$$

where the first equation computes a bit-string that is 1 at positions where the output results differ and the second equation computes the amount of that bits.

4.2.1.2 Accumulation

The accumulation circuit A has state and can store intermediate results. A straightforward implementation is to add all errors:

$$a' = a' + e$$
$$a = a', \tag{A1}$$

where a' is a state variable that stores the current sum of all errors and is initialized with 0. In the remainder of this section, all state variables are primed and initialized with 0. An accumulator can also track the maximum error:

$$a' = \max(a', e)$$
$$a = a'. \tag{A2}$$

When dealing with error-rate and bit masks, binary operations are of interest. As an example to accumulate changed bits one can use the following implementation:

$$a' = a' \mid e$$
$$a = a',$$
(A3)

where "|" refers to bitwise OR. Additional state variables can be employed for more complex computations. In order to keep track of an average error, also the number of cycles needs to be tracked:

$$a' = a' + e$$
$$c' = c' + 1$$
$$a = a'/c'$$
(A4)

4.2.1.3 Decision

Typically the decision circuit implements a comparison with respect to a threshold X, the most common one being

$$\text{bad} = a \geq X,$$
(D1)

that asserts "bad" if the accumulated error is greater or equal than the given threshold.

4.3 Approximation Questions

The approximation miter with the previous discussed implementations for its components can now be combined in several different ways that allows to answer major questions, of which five are presented in the following.

4.3.1 Question 1: What Is the Earliest Time That One Can Exceed an Accumulated Worst-Case Error of X?

This question can be answered by using the implementations (E1), (A1), and (D1) and then performing BMC. If BMC returns a counter-example at time step t, one tries to find another counter-example up to time step $t - 1$.[2] This procedure is repeated until BMC cannot find a counter-example anymore. The earliest time is the latest found t.

4.3.2 Question 2: What Is the Maximum Worst-Case Error?

This question is not a decision problem but an optimization problem. For this purpose we use the approximation miter with implementations (E1), (A2), and (D1). The *worst-case error* is found out using binary search with PDR. The binary search is shown in Algorithm 4.1. X is set to one half of $2^m - 1$ in the first loop, with lower bound 0 and upper bound $2^m - 1$ where m is the bit-width of the considered output vector. PDR is used to solve the miter and if PDR returns True, the lower bound is set to X, otherwise the upper bound is set to $X - 1$. The upper and lower bounds are refined iteratively until they converge to the *worst-case error*.[3]

Algorithm 4.1 Finding maximum worst-case error

1: **function** FIND_WORST_CASE_ERROR
2:     $lower\_bound \leftarrow 0$
3:     $upper\_bound \leftarrow 2^m - 1$
4:     **while** $lower\_bound < upper\_bound$ **do**
5:         $X = \left\lceil \dfrac{(upper\_bound + lower\_bound)}{2} \right\rceil$
6:         $status = PDR(Miter(E1, A2, D1), X)$
7:         **if** $status = $ True **then**
8:             $lower\_bound = X$
9:         **else**
10:             $upper\_bound = X - 1$
11:         **end if**
12:     **end while**
13:     **return** $lower\_bound$
14: **end function**

---

[2]This 2nd check may be required for advanced BMC implementations which do not guarantee a counter-example of minimal length.

[3]Note: The underlying approach using binary search can also be used in combinational circuits. The only differences are to replace the PDR with a conventional SAT solver and the problem formulation steps. A concrete example for such an algorithm is provided later in Chap. 5 on synthesis (Algorithm 5.3).

4.3.3   Question 3: What Is the Earliest Time That One Can Reach an Accumulated Bit-Flip Error of X?

This question can be solved in the same way as Question 1, but with (E2), (A3), and a decision circuit that takes into consideration the number of bits in the accumulated error:

$$\text{bad} = \sum_{i=1}^{m} a_i \geq X$$

4.3.4   Question 4: What Is the Maximum Bit-Flip Error?

This question can be solved in the same way as Question 2 by only replacing (E1) with (E3), in the error computation. The initial lower bound remains the same but the upper bound is initialized to m, the bit-width of the output vector. We use similar binary search as given in Algorithm 4.1 with the PDR call on the *Miter* (E3), (A2), (D1).

4.3.5   Question 5: Can One Guarantee That the Average-Case Error Does Not Exceed X?

Sometimes, one is interested that the error does not grow too large over time although some exceptions are tolerable. As an example, if only small errors have been made for a long time one can accept a larger one. For this purpose, one needs to track the average case error which is done by using the approximation miter with (E1), (A4), and (D1). The result is a decision problem, for which we have: If no counter-example can be found, the average-case behavior can be guaranteed.

The proposed sequential approximation miter is generic and can be configured such that the different major questions can be formulated. By reducing the respective problems either as decision or optimization problem they can be answered using model checking techniques. In the following section these approaches are demonstrated on several examples.

4.4   Experimental Results

All the algorithms are implemented in the *aXc* framework for approximate computing. Several real circuits have a signal to indicate the availability of proper output. This signal, commonly referred to as *output-enable*, is *AND*ed with the corresponding inputs of the miter. Similarly, the value of the output function depends on whether the concerned quantity is signed or unsigned. This has to be additionally indicated to the *aXc* tool with extra options. By default, the program considers all outputs as unsigned. The tool has options to compute the combinational error metrics for comparison purposes. Furthermore, the tool also writes out the counter-example when the *bad* goes high, as a simulation vector input used in test benches for functional verification. This facilitates additional design debugging with functional verification tools.

In the experimental evaluation, the Questions 1–4 from Sect. 4.3 are answered. Question 5 involves the use of a divider in (A4) and the divider is hard to verify formally [HABS14]. The initial experiments on Question 5 did not conclude on practical designs and as a result the implementation of Question 5 is left out for future work. Several publicly available circuits including those from OpenCores and GitHub are used to evaluate the approach. The experiments are carried out on an Octa-Core Intel Xeon CPU with 3.40 GHz and 32 GB memory running Linux 4.1.6. The experimental evaluation is described in two parts. First, an extensive case study of approximated adders in sequential multipliers is discussed in Sect. 4.4.1. Second, the generality and scalability of the approach is demonstrated by applying it to various designs in Sect. 4.4.2.

In the previous chapter, architecturally approximated ad-hoc adder designs such as ACA adders, GDA adders, and GeAr adders were used to demonstrate the techniques for combinational error metric verification (see Sect. 3.5 and Tables 3.1, 3.2 for details). The study of bounds of error metrics for these adder architectures by itself is interesting. However, a more significant information is their impact within a circuit. The isolated combinational behavior of an approximate adder could be very different from when used as a building block in a bigger (sequential) circuit. In the next section, the case study of an 8-bit sequential multiplier using these approximate adder architectures is presented as a motivation. Although these circuits are simple to construct, their error analysis reveals the need and usefulness of the proposed approach.

4.4.1 Approximated Sequential Multiplier

The design used in this experiment is an unsigned 8-bit sequential multiplier with 4-bit inputs. Partial products are computed in each clock cycle and added using an 8-bit adder. The number of clock cycles taken to complete the multiplication is input data dependent with extra checking for the operands being 1 or 0. The adder used in the circuit is instantiated with the published approximation adders from the repository [GA15]. Several adders are chosen from this repository with similar approximation architectures and different configurations. While the design of the non-approximated multiplier is straightforward, several interesting results can be deduced from the approximate versions.

The approximated multiplier designs are evaluated for the questions presented in Sect. 4.3. The results are given in the upper half of Table 4.1. In this table, the design

Table 4.1 Evaluation of Questions in Sect. 4.3 for various designs

| Design details | | | | | | Questions asked | | | |
Circuit	Approximation architecture used	Area*,†	Delay* (ns)	Gates*	Regs*	Q1 (X, t)	Q2	Q3 (X, t)	Q4
Multiplier	*Almost Correct Adder*								
	ACA_II_N8_Q4	507	9.60	205	39	(1000, 41)	128	(5, 11)	3
	ACA_I_N8_Q5	530	11.20	222	39	(1000, 0)	0	(3, 0)	0
	Gracefully Degrading Adder								
	GDA_St_N8_M4_P2	507	9.60	205	39	(1000, 41)	128	(5, 11)	3
	GDA_St_N8_M4_P4	540	10.30	219	39	(1000, 0)	0	(3, 0)	0
	GDA_St_N8_M8_P1	526	7.40	205	39	(1000, 26)	224	(6, 11)	4
	GDA_St_N8_M8_P2	578	8.00	239	39	(1000, 36)	144	(5, 11)	3
	GDA_St_N8_M8_P3	588	8.80	241	39	(1000, 36)	160	(4, 11)	2
	GDA_St_N8_M8_P4	569	9.40	229	39	(1000, 0)	0	(3, 0)	0
	GDA_St_N8_M8_P5	614	9.10	244	39	(1000, 0)	0	(3, 0)	0
	GDA_St_N8_M8_P6	609	10.30	248	39	(1000, 0)	0	(3, 0)	0
	Accuracy Configurable Adder								
	GeAr_N8_R1_P1	526	7.40	205	39	(1000, 26)	224	(6, 11)	4
	GeAr_N8_R1_P2	578	8.00	239	39	(1000, 36)	144	(5, 11)	3
	GeAr_N8_R1_P3	532	9.60	220	39	(1000, 36)	160	(4, 11)	2
	GeAr_N8_R1_P4	530	11.20	222	39	(1000, 0)	0	(3, 0)	0
	GeAr_N8_R1_P5	551	11.20	229	39	(1000, 0)	0	(3, 0)	0
	GeAr_N8_R1_P6	539	11.20	223	39	(1000, 0)	0	(3, 0)	0
	GeAr_N8_R2_P2	507	9.60	205	39	(1000, 41)	128	(5, 11)	3
	GeAr_N8_R2_P4	530	11.20	219	39	(1000, 0)	0	(3, 0)	0

(continued)

Table 4.1 (continued)

Design details						Questions asked			
Circuit	Approximation architecture used	Area*,†	Delay* (ns)	Gates*	Regs*	Q1 (X, t)	Q2	Q3 (X, t)	Q4
	Ripple Carry Adder								
RCA_N8		526	12.00	217	39	(1000, 0)	0	(3, 0)	0
IIR filter‡	Smaller MAC unit	2292	28.30	912	168	(50,000, 19)	31,542	(16, 33)	11
FIR filter◇	Modified coefficients	2149	38.49	827	64	(100,000, 3)	65,536	(16, 2)	16
Quantizer±	Quant. value approx.	4984	7.30	2078	568	(10,000, 40)	274	(9, 5)	9
Stepper Motor	Approx. counter	1479	14.2	658	115	(100, 30)	11	(4, 18)	4
Binary-BCD∓	Approx. conversion	829	7.60	360	61	(200, 41)	8	(4, 17)	4

Questions:
Q1: What is the earliest time (t) that I can exceed an accumulated worst-case error of X?
Q2: What is the maximum worst-case error?
Q3: What is the earliest time (t) that I can reach an accumulated error-rate of X?
Q4: What is the maximum bit-flip error?

*As reported by ABC [MCBJ08] with synthesis command *resyn* and library *mcnc.genlib*
†ABC reports area normalized to *INVX1*
‡IIR Filter is *Direct-form-II Transposed* with 16-bit output
∓16-bit binary input to 5 × 4-bit BCD converter.   Stepper motor output is 8-bit
◇FIR Filter parameters $N = 16$, $M = 17$ ± Design has 8 × 11-bit outputs. Only one verified

details such as *Area*, *Delay*, *Gates*, and *Regs* are taken from the corresponding synthesis run of ABC [MCBJ08] with the command *resyn*. *Delay* is the worst timing delay achieved and *Regs* is the number of sequential elements in the design. The next four columns are the results of the *Questions* presented in Sect. 4.3. In Q1 and Q3, *X* is the error assumed and *t* is the corresponding clock cycles determined by the tool. The last entry in the upper half of the table is the multiplier using a *Ripple Carry Adder* which is the non-approximated reference design.

Some multipliers with approximation adder architectures such as *ACA_I_N8_Q5* and *GDA_St_N8_M8_P4* are equivalent to the non-approximated multiplier. In this case, the designer can safely choose any of them for the given application. This differs drastically from the error behavior of the individual adders given in Table 3.1.

Another interesting aspect is the maximum bit-flip error of the circuits (result of Q4 in Table 4.1). For example, even though the multiplier using the adder *GeAr_N8_R1_P3* has a faster error accumulation (result of Q1) compared to the multiplier with adder *GeAr_N8_R2_P2* (36 cycles are less than 41 which makes it faster), the maximum bit-flip error possible (result of Q4) is better with the former one. This is useful in designing circuits such as the one used for *error detection and correction* since such circuits rely on the error-rate rather than the magnitude of the error introduced. Multipliers using the adders *GDA_St_N8_M8_P1* and *GeAr_N8_R1_P1* have the worst accumulation tendency (result of Q2) and this correlates with their combinational error behavior.

4.4.2 Generality and Scalability

To demonstrate generality and scalability of our proposed approach, various further benchmarks and approximation scenarios are provided in the lower half of Table 4.1. The first design is a 16-bit IIR filter in which the *Multiply and Accumulate* (MAC) unit is replaced by a smaller one. The coefficients and the MAC unit of the original filter are 18-bit wide and the design is a *Direct Form-II Transposed* IIR filter. In the approximated version, the width of the MAC unit is simply reduced to 17-bit. The filter has a maximum worst-case error (result of Q2) of 31,542, which is less than one half of the maximum output value of $2^{15} - 1$ (since the output represents a signed value). But it takes at least 19 cycles for the error to accumulate and cross 50,000 mark. Furthermore, the maximum number of bit-flips (not necessarily the lower bits of the output) in any given clock cycle for this design is 11.

The second design in the lower half of Table 4.1 is a 17-bit output FIR filter, with approximations introduced in the values of coefficients. In this case, the error introduced is higher with an accumulated error crossing 100,000 within 3 cycles of operation and at most 16 bits among the 17 bits toggle for this design in any clock cycle. Similar interpretations can be obtained from the error behavior of the other designs shown in the table such as Quantizer, Stepper Motor, and Binary to BCD converter.

Table 4.2 Run times for Questions in Sect. 4.3 for various designs

Design details						Run-time comparison			
Circuit	Approximation architecture used	Area*,†	Delay* (ns)	Gates*	Regs*	Q1 (s)	Q2 (s)	Q3 (s)	Q4 (s)
Multiplier	*Almost Correct Adder*								
	ACA_II_N8_Q4	507	9.60	205	39	5.12	7.62	2.72	3.20
	ACA_I_N8_Q5	530	11.20	222	39	14.02	11.27	7.90	4.11
	Gracefully Degrading Adder								
	GDA_St_N8_M4_P2	507	9.60	205	39	3.93	9.26	1.70	3.93
	GDA_St_N8_M4_P4	540	10.30	219	39	21.89	12.56	1.77	4.27
	GDA_St_N8_M8_P1	526	7.40	205	39	0.97	5.22	1.00	4.14
	GDA_St_N8_M8_P2	578	8.00	239	39	1.40	8.50	0.76	3.71
	GDA_St_N8_M8_P3	588	8.80	241	39	2.60	7.77	2.78	5.29
	GDA_St_N8_M8_P4	569	9.40	229	39	18.12	12.92	6.20	4.76
	GDA_St_N8_M8_P5	614	9.10	244	39	17.91	12.24	6.13	6.22
	GDA_St_N8_M8_P6	609	10.30	248	39	15.15	16.57	8.62	4.93
	Accuracy Configurable Adder								
	GeAr_N8_R1_P1	526	7.40	205	39	1.67	5.56	1.52	4.72
	GeAr_N8_R1_P2	578	8.00	239	39	3.50	9.68	2.32	3.68
	GeAr_N8_R1_P3	532	9.60	220	39	3.23	7.92	4.02	5.32
	GeAr_N8_R1_P4	530	11.20	222	39	14.03	12.69	7.89	4.80
	GeAr_N8_R1_P5	551	11.20	229	39	14.39	11.78	7.93	4.75
	GeAr_N8_R1_P6	539	11.20	223	39	11.39	24.02	7.42	4.26
	GeAr_N8_R2_P2	507	9.60	205	39	5.14	11.43	2.72	4.65
	GeAr_N8_R2_P4	530	11.20	219	39	14.64	12.64	7.32	4.28

(continued)

Table 4.2 (continued)

	Ripple Carry Adder								
	RCA_N8	526	12.00	217	39	0	0	0	0
IIR filter‡	Smaller MAC unit	2292	28.30	912	168	2.31	272.05	2.34	231.68
FIR filter◇	Modified coefficients	2149	38.49	827	64	1.01	15.78	6.00	2.00
Quantizer⊧	Quant. value approx.	4984	7.30	2078	568	72.98	17.95	3.01	13.86
Stepper Motor	Approx. counter	1479	14.2	658	115	1.00	87.01	1.01	350.15
Binary to BCD⊧	Approx. conversion	829	7.60	360	61	3.01	12.05	1.01	2.44

Questions:
Q1: What is the earliest time (t) that I can exceed an accumulated worst-case error of X?
Q2: What is the maximum worst-case error?
Q3: What is the earliest time (t) that I can reach an accumulated error-rate of X?
Q4: What is the maximum bit-flip error?
Run times correspond to the evaluation in Table 4.1
⋆ As reported by ABC [MCBJ08] with synthesis command *resyn* and library *mcnc.genlib*

The results confirm the applicability of the proposed approach. The run times in seconds taken by the tool to answer these questions are given separately in Table 4.2. The run times heavily depend on the underlying model checking algorithms. Improvement in model checking therefore has a direct positive effect on the methodology as well.

4.5   Concluding Remarks

This chapter introduced the important formal verification techniques proposed for approximate sequential circuits. The main idea is to formulate the error metric verification problem in terms of a sequential approximation miter circuit and apply the standard and well-established formal verification techniques. Our approach is a fully automated one and is capable of providing several useful insights to the error behavior of such circuits. Hence, our techniques stand out in comparison with the existing semi-automated verification methodologies such as [RRV[+]14, VSK[+]12] that employ a rather restricted error criteria (e.g., [VSK[+]12] uses worst-case error and a very closely related relative-error as error metric).

We conclude the formal verification of approximation circuits here. The next chapter deals with the different synthesis techniques developed for approximation circuits.

Chapter 5
Synthesis Techniques for Approximation Circuits

5.1 Overview

This chapter explains an important step in the approximate computing IC design flow—approximation-aware automated synthesis methodologies. Broadly speaking, automated synthesis is the process of converting a behavioral description of a design into a structural netlist targeted to a production technology. The technology independent behavioral description is typically specified in a high level language such as Verilog or VHDL using higher level language semantics such as Register Transfer Level specification. The synthesis tool parses this input, with optional design constraints such as limits in timing/area/power, and builds an internal Boolean network. Further, several optimization techniques are used to minimize the cost metrics associated with the targeted technology and finally the netlist is written out. The underlying synthesis process remains the same in the context of approximate computing too. However, approximate computing can also improve the efficiency of a circuit in terms of speed and area by relaxing the constraints on computational accuracy. Hence, an approximation synthesis tool is able to provide much better optimization compared to a conventional synthesis tool since the functional equivalence need not be maintained. An outline of the approximation synthesis design flow is shown in Fig. 5.1. It has to be noted that a conventional equivalence checking tool typically employed in the post-synthesis stage will not be useful with an approximated netlist. Rather, one must use special approximation-aware equivalence checking techniques introduced in the previous chapter to prove the equivalence. An automated synthesis technique should read in the behavioral description of the circuit, i.e., the RTL and the error metric specification to come up with the final structural netlist.

Several related automated approximation synthesis approaches have been reported in the past. These include reduction of SOP implementations [SG10], redundancy propagation [SG11], and dedicated three-level circuit construction heuristics [BC14]. The SALSA framework and the recently introduced ASLAN

© Springer Nature Switzerland AG 2019

A. Chandrasekharan et al., *Design Automation Techniques
for Approximation Circuits*, https://doi.org/10.1007/978-3-319-98965-5_5

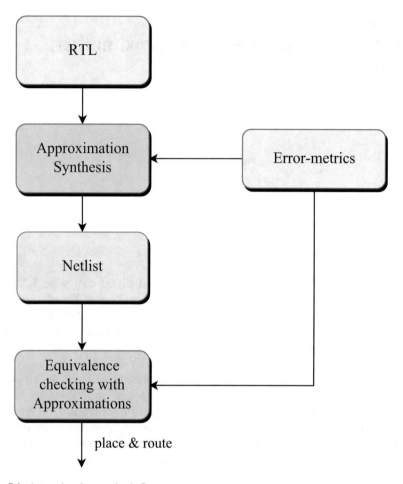

Fig. 5.1 Approximation synthesis flow

extension are able to do efficient synthesis with the help of user defined quality evaluation circuits [VSK⁺12, RRV⁺14]. However, this additional quality evaluation circuit has to be constructed by the user similar to a test bench, which is a design problem by itself. Further, some error metrics such as error-rate cannot be expressed in terms of Boolean functions efficiently since these require counting in the solution space which is a #SAT problem (see Sect. 2.5 in Chap. 2 and Sect. 3.4 in Chap. 3 for details.). Our synthesis techniques do not require any additional test bench circuits. Besides, the different error metrics themselves are not strongly correlated and each application has a varied sensitivity on each of these error metrics. Hence, a synthesis methodology that can take advantage of loosely specified error criteria and adhere to more strictly specified ones provides very good optimization results. Our approaches achieve this, as confirmed by the experimental results provided later in this chapter (see Sect. 5.2.2). These approaches and achieved results have been published in [CSGD16a, SGCD16].

The most important Boolean networks employed in logic synthesis tools are Binary Decision Diagrams and And-Inverter Graphs. As mentioned in the introductory Chap. 1, both these networks are homogeneous, i.e., all the nodes represent the same function. Logic optimization steps are directly related to the data structures they run on. Two different approaches for approximation synthesis are presented in the following. The first one deals exclusively with BDDs and the second one with AIGs. The BDD-based approximation synthesis algorithms cannot guarantee a bound on the resulting network in terms of efficiency (i.e., area or delay). However, for several circuits these are found to be very effective. On the other hand, the AIG-based technique is much more scalable and provides a guarantee on the final result, both on the error behavior and area/delay improvement. First an outline of the BDD-based techniques is presented and later the approximation synthesis using AIG is detailed. Each section covers the algorithms and the respective experimental evaluation. The formal techniques introduced in the previous chapter are used to evaluate the final error characteristics of the resulting network. Note that only combinational circuit synthesis under approximations is considered. Further, only functional approximations are targeted. This means that the resulting synthesized netlist is taken through the later steps such as place-and-route similar to the conventional IC design flow. In line with this, a regular conventional technology mapper is used to map the optimized approximation netlist to the final fabrication technology. In this book, these steps are invoked and the corresponding results are presented only when the differentiation is needed. In general, the comparison and the experimental results obtained in the later sections are provided for the particular network under consideration (i.e., AIG or BDD), rather than the end technology mapped netlist.

5.2   Approximate BDD Minimization

The central concept to BDD-based approximation synthesis is Approximate BDD Minimization (ABM). The *Approximate BDD Minimization* problem aims at minimizing a BDD representation of a given Boolean function f by approximating it with a Boolean function $\hat{f}$ and respecting a given threshold based on an error metric. ABM reduces the *size* of the BDD in terms of the number of nodes.

Let $\hat{f}$ be the approximated Boolean function of f and the procedure $e(f, \hat{f})$ provides the error behavior of $\hat{f}$ wrt. f in terms of the error metrics. Note that this error behavior can be easily found using the techniques introduced in the previous chapter. Further, the procedure $size\_of\_BDD(f)$ provides the size (number of nodes) of the BDD. The associated decision problem for the Approximate BDD Minimization asks for a given function f, limits of error metrics e and a size bound b, whether there exists a function $\hat{f}$ such that $e(f, \hat{f})$ holds and $size\_of\_BDD(\hat{f}) \leq b$. A non-deterministic generic-algorithm is given in Algorithm 5.1 to solve the ABM problem. The details of this algorithm are explained next.

Algorithm 5.1 Approximate BDD minimization

1: **function** APPROX_BDD(BDD f, error behavior e, size limit b)
2:     set $\hat{f} \leftarrow f$
3:     **while** size_of_BDD ($\hat{f}$) > b **do**
4:         set $h \leftarrow APPROX(\hat{f})$
5:         **if** $e(f, h)$ & $size\_of\_BDD(h) < size\_of\_BDD(\hat{f})$ **then**
6:             set $\hat{f} \leftarrow h$
7:         **end if**
8:     **end while**
9:     **return** $\hat{f}$
10: **end function**

Algorithm 5.1 takes in the Boolean function f, the error behavior e, and the size limit for the resulting approximated function b. We use a BDD operator to do the approximations. The function $APPROX$ in Line 3 of the algorithm refers to this BDD operator that approximates $\hat{f}$ further (potentially controlled by user preferences). The details of the approximation BDD operator are explained separately after providing the general details of the algorithm. The result from the $APPROX$ procedure is stored in the temporary variable h. Before $\hat{f}$ can be replaced by h, it needs to be checked whether (1) h has a smaller BDD representation than $\hat{f}$ and (2) the error metric respects the given threshold. The algorithm can be made deterministic by (1) providing a strategy that selects approximation operators in Line 3 and (2) by relaxing the condition in Line 2 to reach the size bound b but terminate already beforehand. The latter condition guarantees that the algorithm completes, however, a heuristic solution may be returned.

Although Algorithm 5.1 is simple and generic, its nontrivial parts are the approximation operator $APPROX$ and the computation of the error metric e directly on the BDD representation. The solution to the first part is presented in the following section. The second part computing the error metrics is covered in the earlier chapter on formal verification of approximate computing (refer Sect. 3.2 for details). For the first part, Sect. 5.2.1 presents five operators that can be used to approximate a function in its BDD representation.

5.2.1   BDD Approximation Operators

This section describes five approximation operators which are summarized in Table 5.1. Applying the operator to a multi-output function denotes applying it to each sub-function. The rounding operators will be described in terms of the $APPLY$ BDD operation from Bryant [Bry86] using the notation proposed by Knuth [Knu11]. We refer to a vertex in the BDD and to the function f it represents interchangeably and make use of the notation

$$f = (\bar{x}_v \ ? \ f_l : f_h) \tag{5.1}$$

Table 5.1 Approximation operators

Operator	Description	Operator	Description
f_{x_i}	Positive co-factor	$\lfloor f \rfloor_{x_i}$	Rounding down
$f_{\bar{x}_i}$	Negative co-factor	$\lceil f \rceil_{x_i}$	Rounding up
		$\lceil f \rfloor_{x_i}$	Rounding

where x_v is the vertex and f_l, f_h refer to the low and high successor, respectively. The efficiency in the APPLY algorithm is mainly due to memoization and the use of a unique table. For the latter, the algorithmic descriptions of the operators will make use of an operation UNIQUE(v, r_l, r_h). It refers to the procedure of looking up whether there already exists a vertex labeled x_v with low successor r_l and high successor r_h and returning it in this case, or creating a new vertex otherwise.

5.2.1.1 Co-Factor Approximation

One of the simplest approximation operators is taking the co-factor with respect to some variable x_i, i.e.,

$$\hat{f} \leftarrow f_{x_i} \quad \text{or} \quad \hat{f} \leftarrow f_{\bar{x}_i}. \tag{5.2}$$

The implementation in a BDD package in terms of the APPLY operation is as follows:

$$f_{x_i} = \begin{cases} \text{Represent } f \text{ as in (5.1).} \\ \text{If } f \text{ is constant or if } v > i, \text{ return } f. \\ \text{Otherwise, if } `f_{x_i} = r' \text{ is in the memo cache, return } r. \\ \text{Otherwise, if } v = i, \text{ set } r \leftarrow f_h. \\ \text{Otherwise, compute } r_l \leftarrow (f_l)_{x_i} \text{ and } r_h \leftarrow (f_h)_{x_i} \\ \quad \text{and set } r \leftarrow \text{UNIQUE}(v, r_l, r_h). \\ \text{Put } `f_{x_i} = r' \text{ into the memo cache, and return } r. \end{cases}$$

The algorithm implements $f_{\bar{x}_i}$ when changing $r \leftarrow f_h$ to $r \leftarrow f_l$ in the fourth step and replacing all occurrences of "$[\cdot]_{x_i}$" by "$[\cdot]_{\bar{x}_i}$."

5.2.1.2 Approximation by Rounding

Two operators are defined $\lfloor f \rfloor_{x_i}$ and $\lceil f \rceil_{x_i}$ for *rounding up* and *rounding down* a function based on the BDD. The idea is inspired by Ravi and Somenzi [RS95]: for each node that appears at level x_i or lower (in other words for each node labeled x_j with $j \geq i$), the lighter child, i.e., the child with the smaller ON-set, is replaced by a terminal node. The terminal node is $\bot$ when rounding down, and $\top$ when rounding up. The technique is called *heavy branch subsetting* in [RS95].

Its algorithmic description based on the APPLY algorithm reads as follows:

$$\lfloor f \rfloor_{x_i} = \begin{cases} \text{If } f \text{ is constant, return } f. \\ \text{Otherwise, if } `\lfloor f \rfloor_{x_i} = r\text{' is in the memo cache,} \\ \quad \text{return } r. \\ \text{Otherwise represent } f \text{ as in (5.1);} \\ \text{If } v < i, \text{ compute } r_l \leftarrow \lfloor f_l \rfloor_{x_i} \text{ and } r_h \leftarrow \lfloor f_h \rfloor_{x_i}; \\ \text{Otherwise, if } |f_l| < |f_h|, \text{ set } r_l \leftarrow \perp \text{ and compute} \\ \quad r_h \leftarrow \lfloor f_h \rfloor_{x_i}; \\ \text{Otherwise compute } r_l \leftarrow \lfloor f_l \rfloor_{x_i} \text{ and set } r_h \leftarrow \perp; \\ \text{Set } r \leftarrow \text{UNIQUE}(v, r_l, r_h); \\ \text{Put } `\lfloor f \rfloor_{x_i} = r\text{' into the memo cache, and return } r. \end{cases}$$

The implementation of $\lceil f \rceil_{x_i}$ equals the one of $\lfloor f \rfloor_{x_i}$ after replacing all occurrences of "$\lfloor \cdot \rfloor$" with "$\lceil \cdot \rceil$" as well as the two occurrences of "$\perp$" with "$\top$".

Figure 5.2a shows a BDD for a function with four inputs and three outputs which serves as an illustrative example for rounding operators. Each example applies rounding at level 3 and for rounding up and rounding down, crosses emphasize lighter children. Figure 5.2b and c shows the resulting BDDs after applying rounding down and rounding up, respectively.

The algorithms for rounding down and rounding up do not necessarily reduce the number of variables since only one child is replaced by a terminal node. The last approximation operator *rounding* does guarantee a reduction of the number of variables since it replaces all nodes of a given level by a terminal node. Which terminal node is chosen depends on the size of the ON-set of the function represented by that node. If the size of the ON-set ($|f|$) exceeds the size of the OFF-set ($|\bar{f}|$), the node is replaced by $\top$, otherwise by $\perp$.

The algorithmic description reads as follows:

$$[f]_{x_i} = \begin{cases} \text{If } f \text{ is constant, return } f. \\ \text{Otherwise, if } `[f]_{x_i}\text{' is in the memo cache, return } r. \\ \text{Otherwise represent } f \text{ as in (5.1);} \\ \text{If } v \geq i \text{ and } |f| > |\bar{f}|, \text{ set } r \leftarrow \top; \\ \text{Otherwise, if } v \geq i \text{ and } |f| \leq |\bar{f}|, \text{ set } r \leftarrow \perp; \\ \text{Otherwise compute } r_l \leftarrow [f_l]_{x_i} \text{ and } r_h \leftarrow [f_h]_{x_i}, \\ \quad \text{and set } r \leftarrow \text{UNIQUE}(v, r_l, r_h); \\ \text{Put } `[f]_{x_i} = r\text{' into the memo cache, and return } r. \end{cases}$$

Figure 5.2d shows the effect of rounding at level 3.

5.2.2 Experimental Evaluation

The evaluation of the BDD-based approximation techniques uses the ISCAS-85 benchmark set. Details of this benchmark are introduced earlier in Chap. 3 (cf. Table 3.3). The experimental evaluation is mainly focused on investigating

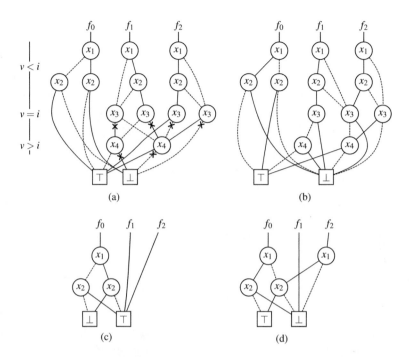

Fig. 5.2 Example for approximation operators. (**a**) Rounding example f. (**b**) Rounding down $\lfloor f \rfloor_{x_3}$. (**c**) Rounding up $\lceil f \rceil_{x_3}$. (**d**) Rounding $[f]_{x_3}$

how the BDD size evolves when increasing the number of levels in the rounding down, rounding up, and rounding operators. Since the co-factor operators consider one level and do not directly effect the successive ones, they are not part of the evaluation. Further only the error-rate metric is considered for Approximate BDD Minimization problem.

The plots in Fig. 5.3 show the results of this evaluation. The x-axis marks the error-rate and the y-axis marks the size improvement of the BDD representation for a particular configuration. The color refers to the approximation operator and a small number above the mark reveals the value for i, i.e., the level at which the operator was applied.

A steep curve means that a high size improvement is obtained by only a small increase in error-rate. A flat curve means the opposite: the error-rate increases significantly by reducing the BDD only slightly. The circuits "c17," "c432," and "c3540" show neither a steep nor a flat curve. In other words, by rounding more parts of the BDD the size can be reduced by accepting a reasonable increase in the error-rate. In "c1908" the curve is very steep at first and then becomes flat, at least for rounding up and rounding operators. A good trade-off is obtained at an error-rate of about 28% and a size improvement of about 92%. The benchmarks "c499" and "c1355" show similar (but not as strong) effects. Also it can be noticed that the effects are not as high for rounding down, which gives a more fine grained control over the approximation.

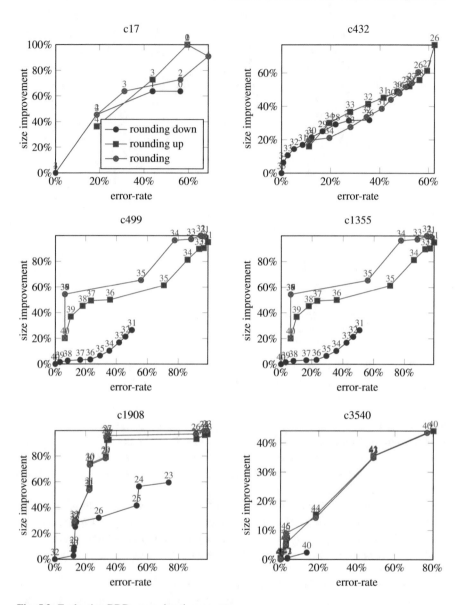

Fig. 5.3 Evaluating BDD approximation operators

The experimental results obtained with the rounding operators show that the BDD approximation is a viable option for small circuits. However, the number of nodes of the BDD correlates rather loosely with the size of the end circuit. Besides, rounding a BDD is not strictly guaranteed to have a reduction in BDD size. It can happen that after removing a node, the resulting BDD size can in fact increase. Optimization techniques based on other Boolean networks such as And-Inverter

Graphs are far more scalable in this regard. Besides, the procedure provided in Algorithm 5.1 needs to compute the error metric in each loop iteration. As explained in Chap. 3, AIG and the associated SAT-based techniques are much more scalable in this aspect also. The remaining part of this chapter focuses on the AIG-based approximation synthesis.

5.3 AIG-Based Approximation Synthesis

In this section, an AIG algorithm for the synthesis of approximation circuits with formal guarantees on error metrics is proposed. Central to the approximation synthesis problem is the *approximation-aware AIG rewriting* algorithm. This technique is able to synthesize circuits with significantly improved performance within the allowed error bounds. It also allows to trade off the relative significance of each error metric for a particular application, to improve the quality of the synthesized circuits. First on this section, a brief review on the relevant terminology for an AIG is provided. Further, the details of the algorithm are provided, followed by experimental results. Experimental evaluation is carried out on a broad range of applications. Our synthesis results are even comparable to the manually hand-crafted approximate designs. In particular, the benefits of approximation synthesis are demonstrated in an image processing application towards the end of this chapter.

5.3.1 And-Inverter Graph Rewriting

As mentioned in Chap. 2, an AIG is a Boolean network where the nodes represent two-input ANDs and the edges can be complemented, i.e., inverted. A *path* in an AIG is a set of nodes starting from a primary input or a constant, and ending at a primary output. The *depth* of an AIG is the maximum length among all the paths and the *size* is the total number of nodes in the AIG. The depth of an AIG corresponds to delay and its size corresponds to the area of the network. The aim of a generic synthesis approach is to reduce the depth and area of the AIG.

Rewriting is an algorithmic transformation in an AIG that introduces local modifications to the network to reduce the depth and/or the size of the AIG [MCB06]. Rewriting takes a greedy approach by iteratively selecting subgraphs rooted at a node and substituting them with better pre-computed subgraphs. We use *cuts* to do the rewriting for AIG networks efficiently. Recall from Chap. 2 (Sect. 2.2.2) that the cut size which is the number of nodes in the transitive fan-in cone is a measure of area of the cut. Further, each k-feasible cut (i.e., number of leaves is less than or equal to k) has a local cut function which is expressed in terms of the leaves as inputs. A k-feasible cut represents a single output function g, with k inputs which may be shared or substituted with another function, $\hat{g}$. For rule-based synthesis rewriting, the substituted function is an *equivalent* function conforming

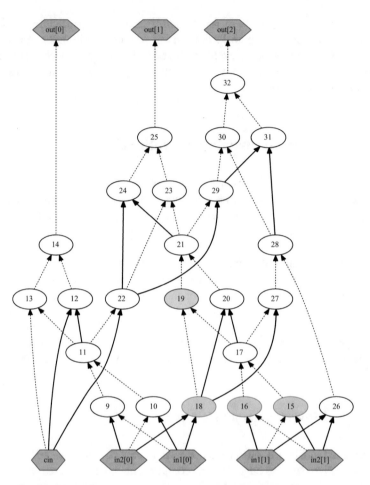

Fig. 5.4 AIG full adder with cut

to the desired synthesis goals [MCB06, Een07, LD11]. However, in approximate synthesis, the substituted function does not have to be equivalent, but should respect the global error and quality metrics, and synthesis goals. Cuts in an AIG can be computed using *cut enumeration* techniques [PL98].

An AIG network of a 2-bit *full adder* circuit is illustrated in Fig. 5.4. Each node other than the terminal nodes (PIs and POs) represents an AND gate, and the dotted arrows indicate inversion of the respective input. A 3-input cut is shown with root node *19* and leaves *18, 16*, and *15*. The size of this cut is 2 (nodes *19, 17*).

5.3.2 Approximation-Aware Rewriting

Our approach applies network rewriting which allows to change the functionality of the circuit, but does not allow to violate a given error behavior. The error behavior is given in terms of thresholds to error metrics. It is possible that a combination of several error metrics is given.

Algorithm 5.2 Approximation rewriting

```
 1: function APPROX_REWRITE(AIG f, error behavior e)
 2:     set f̂ ← f
 3:     while continue do
 4:         set paths ← select_paths(f̂)
 5:         for each p ∈ paths do
 6:             set cuts ← select_cuts(p)
 7:             for each C ∈ cuts do
 8:                 set f̂cnd ← replace C by Ĉ in f̂
 9:                 if e(f, f̂cnd) then
10:                     set f̂ ← f̂cnd
11:                 end if
12:             end for
13:         end for
14:     end while
15:     return f̂
16: end function
```

The rewriting algorithm is outlined in Algorithm 5.2. The description is generic and details on the important steps are described in the next section. The input is an AIG that represents some function f. It returns an AIG that represents an approximated function $\hat{f}$, which complies to the given error behavior e. In the algorithm, we model the error behavior as a function that takes f and $\hat{f}$ as inputs and returns 0, if the error behavior is violated. As an example, we can define the error behavior

$$e(f, \hat{f}) = e_{\mathrm{wc}}(f, \hat{f}) \leq 1000 \wedge e_{\mathrm{bf}}(f, \hat{f}) \leq 5$$

in which the worst-case error should be less than 1000 and the maximum bit-flip error should be less than 5.

The algorithm initially sets $\hat{f}$ to f (Line 2). It then selects paths in the circuit to which rewriting should be applied (Line 4). Cuts are selected from the nodes along these paths (Line 6). For each of these cuts C, an approximation $\hat{C}$ is generated, and inserted as a replacement for C. The result of this replacement is temporarily stored in the candidate $\hat{f}_{\mathrm{cnd}}$ (Line 8). It is then checked, whether this candidate respects the error behavior. If that is the case, $\hat{f}$ is replaced by the candidate $\hat{f}_{\mathrm{cnd}}$ (Line 10). This process is iterated as long as there is an improvement, based on a user provided limit on number of attempts, or long as given resource limits have not been exhausted (Line 3).

5.3.3 Implementation

In this section, we describe details on how to implement Algorithm 5.2. The crucial parts in the algorithm are (1) which paths are selected, (2) which cuts are selected, (3) how cuts are approximated, and (4) how the error behavior is evaluated.

The approximation rewriting algorithm is an iterative approach and a decision has to be taken on how many iterations need to be run before exiting the routine. This is implemented as effort levels (*high*, *medium*, and *low*) in the tool corresponding to the number of paths selected for approximation. The user has to specify this option. Alternately, the user can also specify the number of attempts tried by the tool.

5.3.3.1   Select Paths

The primary purpose of the proposed approximation techniques is to reduce delay and area of the circuits. In order to reduce delay, we select the critical paths, i.e., the longest paths in the circuit. Replacing cuts on these paths with approximated cuts of smaller depth reduces the overall depth of the circuit. In our current implementation of the *aXc* package, we select all critical paths. The set of critical paths changes in each iteration.

5.3.3.2   Select Cuts

While selecting critical paths potentially allows to reduce the depth of the approximated circuit, selecting cuts allows to reduce area. We select cuts by performing cut enumeration on the selected paths. In our implementation the enumerated cuts are sorted based on the increasing order of cut size. The rational for approximating the cuts based on increasing order of cut size is as follows. For a given path in the AIG, if we assume each node has equal probability in inducing errors, the size of the cut can be related to the perturbations introduced in the network and therefore, the cut with the smallest size has the least impact. Hence, starting with a transformation that introduces minimum errors has the best chance of introducing approximations without violating error metrics and falling into a local minima quickly. Although this assumption appears oversimplified, the error metrics (worst-case error, bit-flip error, and error-rate) are independent quantities and do not necessarily correlate with each other. Hence, a quick and efficient way to prioritize cuts is based on cut size going along with the assumption. Our experimental results also confirm the applicability of such an approach. This is the default behavior of the tool. We have tried experiments with maximum cut size first, but this scheme is observed to be falling into local minima at a faster rate and the results are inferior. This further confirms that prioritizing cuts based on increasing order of size, per path, is the most acceptable way. Sometimes, selecting cuts randomly for approximation benefits the rewriting procedure. In the current implementation of the *aXc* tool, this behavior can be optionally enabled by the user.

Fig. 5.5 Approximation
Miter

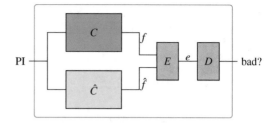

5.3.3.3 Approximate Cut

Each cut is replaced by an approximated cut to generate a candidate for the next approximated network. Ideally, one would like to approximate the cut with a similar cut of better performance, i.e., the error of the function of the approximated cut has minimal errors wrt. to the original cut function, but maximal savings in area and delay. In our current implementation, we simply replace the cut by constant 0, i.e., the root node of the cut is replaced by the constant 0 node. This trivial routine is found to be sufficient for good overall improvements in our experimental evaluation. Investigating how to approximate cuts in a nontrivial manner is a potential area of future research to gain further improvement.

5.3.3.4 Evaluate Error

To compute whether the error behavior is respected, we need a way to precisely compute the error metrics. For this purpose, we make use of an *approximation miter*. An approximation miter takes as input two networks C and $\hat{C}$, an error computation network E, and a decision circuit D. The output of the miter is a single bit *bad* which evaluates to 1 if and only if the error is violated. The general configuration of an approximation miter for combinational networks is illustrated in Fig. 5.5. This directly follows the concepts explained earlier in Chap. 3.

The error computation network E and the decision network D can be configured to do the error analysis after applying approximation rewriting to the AIG. In this work, the worst-case error and bit-flip error are evaluated using the approximation miter. The evaluation of error-rate involves the counting of solutions in $\hat{f}$ that differ from f. We use the algorithms and techniques explained previously in Chap. 3 for error metric computation.

The error metrics e_{wc} and e_{er} can be precisely computed for combinational circuits with the symbolic algorithms explained earlier (see Sects. 3.2 and 3.3 in Chap. 3 for further details). An alternate algorithm to compute the e_{bf} is outlined in Listing 5.3. It is formulated as an optimization problem using an approximation miter and computed with binary search and SAT. This binary search approach is the main differentiation from Algorithm 3.4 provided in Chap. 3. For a function f with output width m, X is set to one half of m in the first loop, with lower bound 0 and

---

Algorithm 5.3 Finding maximum bit-flip error

1: **function** FIND_MAX_BIT_FLIP_ERROR
2:     lbound ← 0
3:     ubound ← $m - 1$
4:     **while** lbound < ubound **do**
5:         $X \leftarrow \left\lceil \dfrac{(\text{ubound} + \text{lbound})}{2} \right\rceil$
6:         $s \leftarrow \text{SAT} \left(\text{ApproxMiter} \left(\sum_{i=0}^{m-1} \left(f_i \oplus \hat{f}_i \right), X \right) \right)$
7:         **if** $s = satisfiable$ **then**
8:             lbound ← X
9:         **else**
10:             ubound ← $X - 1$
11:         **end if**
12:     **end while**
13:     **return** lbound
14: **end function**

---

upper bound $m - 1$. SAT is used to solve the approximation miter and if SAT returns *satisfiable*, the lower bound is set to X, else the upper bound is set to $X - 1$. The binary search algorithm iterates further until the bounds converge to e_{bf}.

5.3.4   Experimental Results

We have implemented all algorithms in C++ as part of the *aXc* framework. The program reads Verilog RTL descriptions of the design and writes out the synthesized approximation netlist. The command compile-rewrite is used to generate the approximate synthesis circuits[1] using approximation-aware rewriting. The experiments are carried out on an Octa-Core Intel Xeon CPU with 3.40 GHz and 32 GB memory running Linux 4.1.6.

In this section, we provide the results of two experimental evaluations. First, we compare the quality of approximate adders synthesized with our approach to state-of-the-art manually architectured approximated adders. The usefulness of the presented automated synthesis technique is further studied in the context of image processing applications. Second, we demonstrate the generality and scalability of the approach by applying it to various designs including standard synthesis benchmark networks such as LGSynth91 [Yan91].

---

[1] The compile command also supports other options. Invoke help using compile -h to get a complete list.

5.3.4.1 Approximate Synthesis of Adders

Approximate synthesis is carried out for adder circuits with a high effort level. The results are given in Table 5.2. These are compared with architecturally approximated adder designs from the repository [GA15]. Many of these architectures are specifically hand-crafted to improve the delay of the circuit.

The case study is carried out as follows. The adders from [GA15] are evaluated for worst-case error and bit-flip error, and then synthesis is carried out by specifying these values as limits, hence, the synthesis result obtained from our approach cannot be worse. The error-rate is left unspecified and synthesis is allowed to capitalize on this.

The left side of Table 5.2 lists the error metrics for architecturally approximated adders, evaluated as given in Sect. 5.3.3.4. The performance metrics such as delay and area are compared with the non-approximated *Ripple Carry Adder* (RCA). The same RCA circuit is given as the input to the approximation synthesis tool along with the e_{wc} and e_{bf} achieved with the architecturally approximated schemes. The synthesized circuits are subsequently evaluated for the error metrics to get the achieved synthesis numbers.

For most of the approximation schemes, our synthesis approach is able to generate circuits with a better area and closer delay compared to the architecturally approximated counterparts, at the cost of error-rate. A large number of schemes such as appx2, appx4, appx5, appx8, and appx10 have significantly improved area with delay numbers matching those of architectural schemes.[2] This study demonstrates that our automatic synthesis approach can compete with the quality obtained from hand-crafted architecturally designs (Table 5.3).

5.3.4.2 Image Processing Application

In order to confirm the quality results of the proposed approach, we show their usage in a real-world image compression application. We have used the OpenCores image compression project [Lun16] to study the impact of approximation adders in signal processing. The project implements the JPEG compression standard and the complete source code is available publicly in the url: http://opencores.org/project,jpegencode. Our experimental setup is as follows. The adders in the color space transformation module of the image compression circuit are replaced with the approximation adders synthesized using our approach and some of the architecturally approximated adders. The input image is the well-known standard test image taken from Wikipedia,[3] trimmed to the specific needs of the image compression circuits. The images generated using these circuits are compared

---

[2]This can be seen by a line-by-line comparison.
[3]https://en.wikipedia.org/wiki/Lenna.

Table 5.2 Synthesis comparison for approximation adders

Approximation architecture				Approximation synthesis				
Architecture	Gates*	Delay* (ns)	Area*	Synthesis scheme[†] ($e_{wc\text{-}in}$, $e_{bf\text{-}in}$)	Gates*	Delay* (ns)	Area*	Time (s)
8-bit adders				*8-bit adders*				
RCA_N8[‡]	57	10.2	175	RCA_N8[‡]	57	10.2	175	
ACA_II_N8_Q4[±]	39	7	137	appx1 (64, 5)	41	7	138	50
ACA_I_N8_Q5	52	7	175	appx2 (128, 4)	27	7	86	57
GDA_St_N8_M4_P2[∓]	39	7	137	appx1 (64, 5)	41	7	138	50
GDA_St_N8_M4_P4	37	9	134	appx3 (64, 3)	36	8.6	121	68
GDA_St_N8_M8_P1	26	3.8	108	appx4 (168, 7)	13	3.8	33	11
GDA_St_N8_M8_P2	35	5.4	124	appx5 (144, 6)	15	5.4	45	56
GDA_St_N8_M8_P3	45	7	149	appx6 (128, 5)	19	7	64	22
GDA_St_N8_M8_P4	44	7	157	appx2 (128, 4)	27	7	86	57
GDA_St_N8_M8_P5	63	8	194	appx7 (128, 3)	31	8.6	104	58
GeAr_N8_R1_P1[‡‡]	26	3.8	108	appx4 (168, 7)	13	3.8	33	11
GeAr_N8_R1_P2	35	5.4	124	appx5 (144, 6)	15	5.4	45	56
GeAr_N8_R1_P3	47	7	153	appx6 (128, 5)	19	7	64	22
GeAr_N8_R1_P4	52	7	175	appx2 (128, 4)	27	7	86	57
GeAr_N8_R1_P5	43	8.6	147	appx7 (128, 3)	31	8.6	104	58
GeAr_N8_R2_P2	39	7	137	appx1 (64, 5)	41	7	138	50
GeAr_N8_R2_P4	37	8.6	132	appx3 (64, 3)	36	8.6	121	68
16-bit adders				*16-bit adders*				
RCA_N16[‡]	93	13.4	303	RCA_N16[‡]	93	13.4	303	0
ACA_II_N16_Q4[±]	75	7	269	appx8 (17,472, 13)	41	7	120	151
ACA_II_N16_Q8	104	10.2	331	appx9 (4096, 9)	94	13.4	254	229
ACA_I_N16_Q4	103	7	321	appx10 (34,944, 13)	41	7	120	150
ETAII_N16_Q4[††]	75	7	269	appx8 (17,472, 13)	41	7	120	151
ETAII_N16_Q8	104	10.2	331	appx9 (4096, 9)	94	13.4	254	229
GDA_St_N16_M4_P4[∓]	110	10	358	appx9 (4096, 9)	94	13.4	254	229
GDA_St_N16_M4_P8	119	11.1	381	appx11 (4096, 5)	95	13.4	277	201
GeAr_N16_R2_P4[‡‡]	81	8.6	284	appx12 (16,640, 11)	89	12.7	226	187
GeAr_N16_R4_P4	104	10.2	331	appx9 (4096, 9)	94	13.4	254	229
GeAr_N16_R4_P8	89	11.8	301	appx11 (4096, 5)	95	13.4	277	201
GeAr_N16_R6_P4	114	10.2	375	appx13 (1024, 7)	94	13.4	264	220

*Reported by ABC [MCBJ08] with library *mcnc.genlib*. Area normalized to *INVX1*

[†] $e_{wc\text{-}in}$, $e_{bf\text{-}in}$: error criteria (worst-case error and bit-flip error) are inputs to the tool

[‡], [±], [∓], [‡‡], [††] Abbrev are as given in: http://ces.itec.kit.edu/1025.php [GA15]

[‡] RCA_N8 and RCA_N16 are 8-bit, 16-bit *ripple carry adders* (reference designs)

[±]ACA is *Almost Correct Adder* [KK12], [∓] GDA is *Gracefully Degrading Adder* [YWY+13]

[‡‡]GeAr is *Generic Accuracy Config Add.* [SAHH15], [††] ETA is *Error Tolerant Add.* [ZGY09]

Table 5.3 Error Metrics comparison for approximation adders

Approximation architecture				Approximation synthesis			
Architecture	e_{wc}	e_{er} (%)	e_{bf}	Synthesis scheme[†] (e_{wc-in}, e_{bf-in})	e_{wc}	e_{er} (%)	e_{bf}
8-bit adders				*8-bit adders*			
RCA_N8[‡]	0	0.00	0	RCA_N8[‡]	0	0.00	0
ACA_II_N8_Q4[±]	64	18.75	5	appx1 (64, 5)	64	75.00	4
ACA_I_N8_Q5	128	4.69	4	appx2 (128, 4)	128	78.22	4
GDA_St_N8_M4_P2[∓]	64	18.75	5	appx1 (64, 5)	64	75.00	4
GDA_St_N8_M4_P4	64	2.34	3	appx3 (64, 3)	64	50.00	3
GDA_St_N8_M8_P1	168	60.16	7	appx4 (168, 7)	128	96.94	7
GDA_St_N8_M8_P2	144	30.08	6	appx5 (144, 6)	144	94.75	6
GDA_St_N8_M8_P3	128	12.50	5	appx6 (128, 5)	128	88.33	5
GDA_St_N8_M8_P4	128	4.69	4	appx2 (128, 4)	128	78.22	4
GDA_St_N8_M8_P5	128	1.56	3	appx7 (128, 3)	128	62.70	3
GeAr_N8_R1_P1[‡‡]	168	60.16	7	appx4 (168, 7)	128	96.94	7
GeAr_N8_R1_P2	144	30.08	6	appx5 (144, 6)	144	94.75	6
GeAr_N8_R1_P3	128	12.50	5	appx6 (128, 5)	128	88.33	5
GeAr_N8_R1_P4	128	4.69	4	appx2 (128, 4)	128	78.22	4
GeAr_N8_R1_P5	128	1.56	3	appx7 (128, 3)	128	62.70	3
GeAr_N8_R2_P2	64	18.75	5	appx1 (64, 5)	64	75.00	4
GeAr_N8_R2_P4	64	2.34	3	appx3 (64, 3)	64	50.00	3
16-bit adders				*16-bit adders*			
RCA_N16[‡]	0	0.00	0	RCA_N16[‡]	0	0	0
ACA_II_N16_Q4[±]	17,472	47.79	13	appx8 (17,472, 13)	8320	99.64	13
ACA_II_N16_Q8	4096	5.86	9	appx9 (4096, 9)	2038	99.80	9
ACA_I_N16_Q4	34,944	34.05	13	appx10 (34,944, 13)	8320	99.64	13
ETAII_N16_Q4[††]	17,472	47.79	13	appx8 (17,472, 13)	8320	99.64	13
ETAII_N16_Q8	4096	5.86	9	appx9 (4096, 9)	2038	99.80	9
GDA_St_N16_M4_P4[∓]	4096	5.86	9	appx9 (4096, 9)	2038	99.80	9
GDA_St_N16_M4_P8	4096	0.18	5	appx11 (4096, 5)	496	96.88	5
GeAr_N16_R2_P4[‡‡]	16,640	11.55	11	appx12 (16,640, 11)	4090	99.90	11
GeAr_N16_R4_P4	4096	5.86	9	appx9 (4096, 9)	2038	99.80	9
GeAr_N16_R4_P8	4096	0.18	5	appx11 (4096, 5)	496	96.88	5
GeAr_N16_R6_P4	1024	3.08	7	appx13 (1024, 7)	1024	99.22	7

[†]e_{wc-in}, e_{bf-in}: error criteria (worst-case and bit-flip errors) given as input to the tool
e_{wc}, e_{er}, e_{bf}: worst-case error, error-rate and bit-flip error
‡ ±, ∓, ‡‡, †† Abbrev are as given in: http://ces.itec.kit.edu/1025.php [GA15]
‡RCA_N8 and RCA_N16 are 8-bit, 16-bit *ripple carry adders* (reference designs)
Note: refer Table 5.2 for details on other abbreviations used

with the non-approximated design using ImageMagick.[4] These images are shown in Table 5.4. Only the image obtained with *appx-50k* (an approximate adder synthesized with e_{wc-in} set to 50,000) is heavily distorted. All other generated images may still be considered as of acceptable quality depending on the specific use case. For comparison, we used ACA_II_N16_Q4 and ETAII_N16_Q8 as the architecturally approximated adders.[5] The image quality is comparable to the synthesized approximate adders. Both sets of images do not appear to have a big quality loss despite the high error-rate in approximation synthesis adders. This is due to human perceptual limitations.

A quantitative analysis of the distortions introduced due to approximations can be done using the PSNR (*Peak Signal to Noise Ratio*) plots given in the latter part of Table 5.4. Using the plots, the difference can be better judged. As can be seen, the synthesized adders show comparable measures to the architectural adders.

In this application case study, the approximation adders are used without considering the features and capabilities of the compression algorithm in depth. A detailed study of approximation adders in the context of image processing is given in [SG10, GMP+11, MHGO12].

5.3.4.3   Note on Error-Rate

As can be seen from the results in Table 5.2, the synthesized approximated adders have a higher error-rate. However, this has no effect on the quality in many scenarios, as, e.g., shown in the image compression case study. The error-rate is a metric that relates to the number of errors introduced as a result of approximation. In many signal processing applications involving arithmetic computations (e.g., image compression), designers may choose to focus on other error metrics such as worst-case error [LEN+11]. Since the decision is already taken to introduce approximations, the *impact* or the magnitude of errors could be of more significance than the absolute total number of errors itself. Besides, for a general sequential circuit, errors tend to *accumulate* over a period of operation. Though it may be argued that circuits with higher error-rate have higher chance of accumulating errors, in practice, this is strongly dependent on the *composition* of the circuit itself and the input data. Further details on estimating the impact of errors in sequential circuits have been explained previously in Chap. 4. Nevertheless, there is a broad range of applications where error-rate is an important metric in the design of approximate hardware [Bre04, LEN+11].

---

[4]http://www.imagemagick.org/.

[5]We use the naming convention given in the repository [GA15].

Table 5.4 Image Processing with Approximation Adders

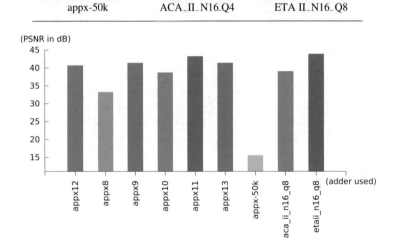

Approximation adder schemes vs PSNR achieved

5.3.4.4 Generality and Scalability

We evaluated our method for a wide range of designs and benchmark circuits. The results given in Table 5.5 show the generality and applicability of our method. A subset of the LGSynth91 [Yan91] circuits are given in the left side of the table. Each circuit is synthesized in three flavors: (1) specifying values of all the error metrics together, (2) specifying only the error-rate, and (3) specifying both worst-case error and bit-flip error, leaving out error-rate. The achieved delay and area in these three schemes are compared with the original non-approximated circuit given as the first entry in each section. In a similar way, several other circuits such

Table 5.5 Approximate synthesis results: LGSynth91 [Yan91]

Design†/synthesis†† ($e_{wc\text{-}in}$, $e_{bf\text{-}in}$, $e_{er\text{-}in}$)††	Gates*	Delay* (ns)	Area*	e_{wc}	e_{er} (%)	e_{bf}	Time (s)
cm163a (I:16,O:5)	34	5.70	78	0	0	0	0
*appx*1 (16, 3, 50)	15	4.10	25	14	43	3	7
*appx*2 (−1, −1, 25)	18	3.00	36	14	21	3	1
*appx*2 (10, 2, −1)	20	4.70	41	8	88	2	4
z4ml (I:7,O:4)	31	12.10	84	0	0	0	0
*appx*1 (8, 1, 75)	20	8.70	52	2	50	1	3
*appx*2 (−1, −1, 25)	31	12.10	84	0	0	0	7
*appx*3 (4, 3, −1)	5	3.80	13.00	4	82	2	3
alu2 (I:10,O:6)	259	32.20	627	0	0	0	0
*appx*1 (30, 3, 50)	236	32.20	570	16	45	2	37
*appx*2 (−1, −1, 25)	231	32.20	566	16	24	1	1
*appx*3 (20, 6, −1)	8	3.30	16	19	81	3	12
frg1 (I:28,O:3)	129	27.10	321	0	0	0	0
*appx*1 (3, 2, 50)	128	27.10	317	1	44	1	10
*appx*2 (−1, −1, 25)	126	27.10	313	2	16	1	1
*appx*2 (2, 1, −1)	128	27.10	317	1	56	1	3
alu4 (I:14,O:8)	519	40.00	1247	0	0	0	0
*appx*1 (128, 4, 50)	489	40.00	1172	64	22	1	139
*appx*2 (−1, −1, 25)	489	40.00	1172	64	22	1	1
*appx*3 (80, 6, −1)	239	34.70	557	79	95	5	55
unreg (I:36,O:16)	83	3.40	227	0	0	0	0
*appx*1 (32,000, 8, 50)	80	3.40	214	512	38	1	45
*appx*2 (−1, −1, 25)	83	3.40	227	0	0	0	1
*appx*3 (10,000, 12, −1)	46	3.40	90	9088	99	10	17
x2 (I:10,O:7)	30	5.80	74	0	0	0	0
*appx*1 (64, 4, 50)	23	5.70	53	64	37	3	7
*appx*2 (−1, −1, 25)	27	5.70	66	64	12	1	1
*appx*3 (50, 6, −1)	17	5.60	41	40	1	3	9
count (I:35,O:16)	120	14.60	261	0	0	0	0
*appx*1 (32,000, 8, 50)	104	14.60	220	7	43	3	140
*appx*2 (−1, −1, 25)	53	3.00	110	65,535	24	16	1
*appx*3 (10,000, 12, −1)	101	14.40	209	8	97	4	141

*Reported by ABC [MCBJ08] with library *mcnc.genlib*. Area normalized to *INVX1*
†Design name given with PI (I) and PO (O) in parenthesis
††$e_{wc\text{-}in}$, $e_{bf\text{-}in}$ and $e_{er\text{-}in}$ (in %) are input error criteria given to *aXc* tool
Value -1 indicates that the particular metric is not enforced. *aXc* effort-level low
Synthesized output circuits are *appx*1, *appx*2 and *appx*3

as multipliers (Array, Wallace tree, and Dadda tree) and multiply accumulators (MACs) are given next. Besides these standard arithmetic designs, other circuits such as parity generator, priority encoder, and BCD converters are also synthesized

Table 5.6 Approximate-synthesis results: other designs

Design[†]/synthesis[††] ($e_{\text{wc-in}}$, $e_{\text{bf-in}}$, $e_{\text{er-in}}$)[††]	Gates[⋆]	Delay[⋆] (ns)	Area[⋆]	e_{wc}	e_{er} (%)	e_{bf}	Time (s)
Multipliers and MAC[‡]							
ArrayMul (I:16,O:16)	420	33.40	1193	0	0	0	0
*appx*1 (32,000,4,50)	420	33.40	1188	128	49	1	759
*appx*2 (−1,−1,25)	435	31.60	1234	256	24	8	1
*appx*3 (20,000,14,−1)	404	33.20	1086	16,448	99	9	107
WallaceMul (I:16,O:16)	398	33.50	1156	0	0	0	0
*appx*1 (32,000, 4, 50)	397	33.50	1146	512	49	1	1055
*appx*2 (−1,−1, 25)	391	31.50	1142	512	23	7	3
*appx*3 (20,000, 14, −1)	352	33.50	1029	5952	99	10	82
DaddaMul (I:16,O:16)	383	30.20	1082	0	0	0	0
*appx*1 (32,000, 4, 50)	382	30.20	1072	1024	47	1	1145
*appx*2 (−1, −1, 25)	368	30.20	1047	832	24	10	7
*appx*3 (20,000, 14, −1)	331	30.20	933	2368	99	10	78
Parity[◇] (I:32,O:36)	136	13.00	276	0	0	0	0
*appx*1 (−1, 1, −1)[**]	111	13.00	215	4G	50	1	29
*appx*2 (−1, 2, −1)[**]	86	13.00	154	12G	75	2	24
*appx*3 (−1, 3, −1)[**]	61	13.00	94	30G	88	3	24
Priority[‡‡] (I:32,O:36)	96	26.30	225	0	0	0	0
*appx*1 (1, −1, −1)	78	19.60	176	1	83	1	168
*appx*2 (4, −1, −1)	45	16.90	91	4	96	3	69
*appx*3 (10, −1, −1)	43	12.10	94	8	40	4	12
Bin2BCD[±] (I:8,O:10)	240	30.20	563	0	0	0	0
*appx*1 (−1, −1, 10)	231	28.00	529	576	6	2	1
*appx*2 (−1, −1, 20)	229	28.00	524	576	19	5	1
*appx*3 (−1, −1, 30)	214	27.50	492	110	28	7	1
BCD2Bin[∓] (I:10,O:8)	64	16.10	209	0	0	0	0
*appx*1 (10, −1, −1)	62	16.10	194	8	9	3	33
*appx*2 (25, −1, −1)	62	16.10	189	22	93	4	31
*appx*3 (50, −1, −1)	61	16.10	182	46	97	5	19

Note: refer Table 5.5 for abbreviations
[⋆]Reported by ABC [MCBJ08] with library *mcnc.genlib*. Area normalized to *INVX1*
[†]Design name given with PI (I) and PO (O) in parenthesis
[††]$e_{\text{wc-in}}$, $e_{\text{bf-in}}$ and $e_{\text{er-in}}$ (in %) are input error criteria given to *aXc* tool
[‡]Multipliers and multiply accumulate (MAC) designs are generated from [Aok16]
[‡]http://www.aoki.ecei.tohoku.ac.jp/arith [Aok16].
[◇]Parity generator (4 bits parity, 32 bits data). [‡‡]32 to 5 priority encoder
[**]G stands for a multiplier of 10^9. Numerical precision omitted for brevity
[±], [∓]Bin to BCD and BCD to Bin converters; 3 digit BCD and 10 bit binary

and the results are given in Table 5.5. In almost all cases, the *aXc* synthesis tool is able to optimize area exploiting the flexibility in the provided error limits. In many cases, delay is also simultaneously optimized along with area. As a consequence, the synthesized approximated circuits have a substantially improved *area-delay* product value. In general, as the circuit size (area and gate count) reduces, the power consumed by the circuit also decreases. Hence, these approximated circuits also benefit from reduced power consumption (Table 5.6).

5.4 Concluding Remarks

In this chapter, we proposed automatic synthesis approaches for approximate circuits. The proposed methodologies have several advantages over the current state-of-the-art techniques such as [VSK$^+$12]. Our method can synthesize high quality approximation circuits within the user-specified error bounds for worst-case error, bit-flip error, and error-rate. Experimental evaluation on several applications confirm that our methodology has a large potential and the synthesized circuits are even comparable to hand-crafted architecturally approximated circuits in quality. Besides, we presented case studies where the ability of our method to significantly improve circuit performance, capitalizing on less significant error criteria while respecting more stringent ones, is demonstrated.

Chapter 6
Post-Production Test Strategies for Approximation Circuits

Post-production test or simply test is the process of sorting out defective chips from the proper ones after fabrication. This chapter examines the impact of approximations in post-production test and proposes test methodologies that have the potential for significant yield improvement. To the best of our knowledge, this is the first systematic approach considering *the impact of design level approximations in post-production test.*

6.1 Overview

A wide range of applications benefit from the approximate computing paradigm significantly. This is primarily due to the *controlled* insertion of functional errors in the design that can be tolerated by the end application. These errors are introduced into the design either manually by the designer or by approximate synthesis approaches. The previous chapter on synthesis is dedicated to the synthesis of such systems using an error specification (see Chap. 5). Further, verification techniques introduced in Chaps. 3 and 4 can be used to formally verify the limits and the impact of the errors in the design. From here, the standard design flow is taken and the final layout data is send to the fab for production. After fabrication, the manufactured approximate computing chip is eventually tested for production errors using well-established fault models. To be precise, if the test for a test pattern fails, the approximate computing chip is sorted out. However, from a general perspective this procedure results in throwing away chips which are perfectly fine taking into account that the considered fault (i.e., physical defect that leads to the error) can still be tolerated because of approximation. This can lead to a significant amount of yield loss. In general, the task of manufacturing test is to detect whether a physical defect is present in the chip or not. If yes, the chip will not be shipped to the customer. However, given an approximate circuit and a physical defect, the crucial question

© Springer Nature Switzerland AG 2019
A. Chandrasekharan et al., *Design Automation Techniques for Approximation Circuits*, https://doi.org/10.1007/978-3-319-98965-5_6

is, whether the chip still can be shipped since the defect can be tolerated under approximation in the end application. If we can provide a positive answer to this question, this leads to a significant potential for yield improvement.

In this chapter, an approximation-aware test methodology is presented. It is based on a pre-process to identify *approximation-redundant* faults. By this, all the potential faults that no longer need to be tested are removed because they can be tolerated under the given error metric. Hence, no test pattern has to be generated for these faults. This test methodology is based on SAT and structural techniques and can guarantee whether a fault can be tolerated under approximation or not. The approach has been published in [CEGD18]. The experimental results and case studies on a wide variety of benchmark circuits show that, depending on the approximation and the error metric (which is driven by the application), a relative reduction of up to 80% in fault count can be achieved. The impact of approximation-aware test is a significant potential for yield improvement.

The approximation-aware test methodology in the wider context of chip design and fabrication is shown in Fig. 6.1. An *Approximation Fault Classifier* that

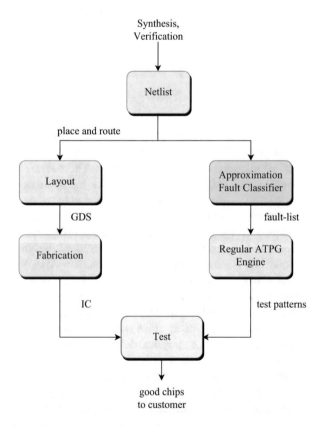

Fig. 6.1 Approximation-aware test and design flow

generates the list of faults targeted for ATPG generation is the main highlight of this scheme. The methodology shown as *Approximation Fault Classifier* in the block diagram does not radically change the test or design flow, rather it is a pre-process to classical Automatic Test Pattern Generation (ATPG) tool. In fact, the only extra addition in classical Design For Test (DFT) is the Approximation Fault Classifier. The key idea is to classify each fault (the logical manifestation of a defect) in *approximation-redundant* or *non-approximation*. For this task, we essentially compare the non-approximated (golden) design against the approximated design with an injected fault under the considered error metric constraint. Using formal methods (SAT and variants) as well as structural techniques allows the fault to be classified into these two categories. After the fault classification, the resulting fault list is given as input to the classical ATPG tool. This fault list contains only the non-approximation faults. Hence, the test pattern is generated only for these faults. Depending on the concrete approximation and error metric (which is driven by the application), a relative reduction of up to 80% in fault count can be achieved. This is demonstrated for a wide range benchmarks provided towards the end of this chapter. This can improve the yield significantly. A brief review of the related work is presented next. This is essential to differentiate our work from similar concepts introduced earlier.

Several works have been proposed to improve yield by classifying faults as *acceptable* faults and *unacceptable* faults (or alternately *benign* and *malign* faults). These employ different techniques such as integer linear programming [SA12], sampling methods for error estimation [LHB05], and threshold-based test generation [ISYI09] . Further, [LHB12] shows a technique to generate tests efficiently if such a classification is available.

However, all these approaches are applied to conventional circuits without taking into consideration the errors introduced as part of the design process itself. Therefore, these approaches cannot be directly applied to approximate computing. It has to be noted that "normal" circuits that produce errors due to manufacturing defects do not constitute approximation circuits. In approximate computing, errors are introduced into the design for high speed or low power. In other words the error is *already* introduced and taken into consideration during design time. Now if for these designed approximated circuits arbitrary fabrication errors are allowed, the error effects will magnify. For instance, if we discard all the stuck-at faults at the lower bit of an approximation adder under a worst-case error constraint of at most 2, the resulting error can in fact increase above the designed limit. Therefore, the end application will fail under such defects. This is exemplified later in a motivating example in Sect. 6.2.1. The key of an approximation-aware test methodology is to identify all the faults which are guaranteed not to violate the given error metric constraint, coming from the end application. This ensures that the approximate computing chip will work as originally envisioned for. At this point, it is important to differentiate this scheme from [WTV+17]. In [WTV+17], structural analysis is used to determine the most vulnerable circuit elements. Only for those elements test patterns are generated and this approach is called *approximate test*. In addition, note that [WTV+17] targets "regular" non-approximated circuits. Hence, this is categorized as a technique for *approximating a test*, rather than a technique for testing an already approximated circuit.

The remainder of this chapter is structured as follows. We introduce the proposed approximation-aware test in the next section, followed by the experimental evaluation of our methodology. The concluding remarks are provided after this experimental evaluation.

6.2   Approximation-Aware Test Methodology

In this section, the approximation-aware test methodology is introduced. Before the details are provided, the general idea using a motivating example is described. In the second half, the proposed fault classification approach is presented.

6.2.1   General Idea and Motivating Example

In the context of approximate computing, yield improvement can be achieved when a fault (logical manifestation of a physical defect) is found which can still be tolerated under the given error metric. In this case, the fabricated chip can still be used as originally intended, instead of sorting it out. In this work, only the stuck-at fault model and stuck-at faults are considered. These are the stuck-at 0 (SA0) and stuck-at 1 (SA1) faults. Given an approximate circuit, a constraint wrt. an error metric and the list of all faults for the approximate circuit, then each fault is categorized by the approximation-aware fault classifier into one of the following:

- *approximation-redundant fault*—These are the faults which can be approximated, i.e., the fault effect can have an observable effect on the outputs, but it is proven that the effect will always be below the given error limit. Hence, no test pattern is needed for these faults. Note that regular redundant faults are also classified into this category.
- *non-approximation fault*—These are faults whose error behavior is above the given error limit. Hence, they have to be tested in the post- production test and thus a test pattern has to be generated for these faults.

Further, if a fault cannot be classified due to reasons of complexity, it is treated as a non-approximation fault.

In the following, a motivating example is provided to demonstrate both fault categories. Consider the 2-bit approximation adder as shown in Fig. 6.2. This adder has two 2-bit inputs $a = a_1a_0$ and $b = b_1b_0$ and the carry input c_{in} and computes the sum as $c_{out}sum_1sum_0$. The (functional) approximation has been performed by cutting the carry from the full adder to the half adder as can be seen in the block diagram on the left of Fig. 6.2. As error metric consider a worst-case error of 2 (coming from the application where the adder is used). Hence, the application can

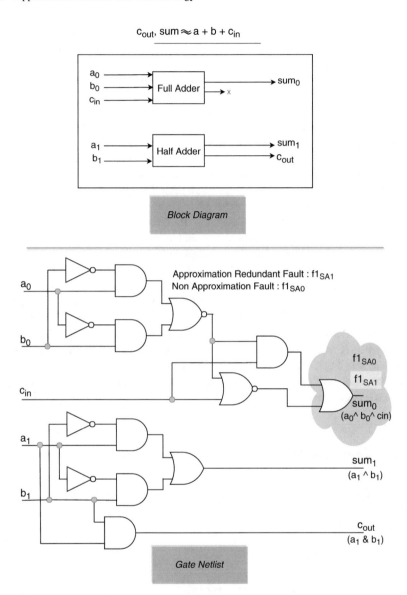

Fig. 6.2 Approximation Adder

tolerate any error magnitude below 2. To explain the proposed fault classification
we will focus on the output bit sum_0 and the faults at this bit, i.e., $f1_{SA0}$ and $f1_{SA1}$
corresponding to a stuck-at-0 and stuck-at-1 fault, respectively.

The truth table of the original golden adder, the approximation adder, and the
approximation adder with different fault manifestations is given at the right side
of Fig. 6.2. The first column of the truth table is the input applied during fault

simulation, followed by the output response of the correct golden adder. Next the response of the approximation adder and the error $e^{\ddagger}$ (as an integer) is shown. The worst-case error $e_{wc}^{\ddagger}$ is the maximum among all such $e^{\ddagger}$. As can be seen the maximum is 2, since cutting the carry leads sometimes to a "wrong" computation but the deviation from the correct result is always less than or equal to 2. The next four columns are the output and error response of the approximation adder with the stuck-at fault, i.e., SA0 and SA1 at the sum_0 output bit. The shaded rows in Table 6.1 correspond to the maximum error in the respective cases and the input pattern corresponding to this error. Recall, since the adder is used for approximate computing applications all the errors below the worst-case error of $e_{wc}^{\ddagger} = 2$ are tolerated. Under this error criteria, the SA1 fault $f1_{SA1}$ at the sum_0 output bit is approximation-redundant because error $e^{\pm}$ is always less than or equal to 2, as can be seen in the rightmost column of the truth table. However, for the same output bit, the SA0 fault $f1_{SA0}$ is a non-approximation fault: the worst-case error is 3 which becomes evident in column $e^{\star}$ and the shaded rows. Hence, for this example circuit, the test need to target the SA0 fault at the sum_0 ($f1_{SA0}$ in Fig. 6.2) whereas it is safe to ignore the SA1 fault ($f1_{SA1}$) in the same signal line.

In practice, the employed error criteria follows the requirements of the approximate computing application. Each application will have a different sensitivity on the error metrics such as worst-case error, bit-flip error, or error-rate. However, if we can identify many approximation-redundant faults, they do not have to be tested since they can be tolerated based on the given error metric constraint.

In the next section, the fault classification algorithm which can handle the different error metrics is presented.

6.2.2   Approximation-Aware Fault Classification

At first, the overall algorithm is presented. Then, the core of the algorithm is detailed.

6.2.2.1   Overall Algorithm

The main part of the proposed approximation-aware fault classification methodology is the fault-preprocessor. It classifies each fault into the above introduced fault categories and is inspired by regular SAT-based ATPG approaches, since these approaches are known to be very effective in proving redundant faults.

The approximation-aware fault classification algorithm is outlined in Algorithm 6.1. The algorithm is generic and details on the individual steps are given below. The inputs are the list of all faults and the error behavior. This error behavior is specified in terms of a constraint wrt. an error metric, e.g., the worst-case error should be less than 10. Such information can be easily provided by the designer of the approximation circuit. Initially the design is parsed and the internal *Netlist* data

Table 6.1 Truth table for approximation adder in Fig. 6.2

	Correct[†] adder	Approx[‡] adder		Appx:SA0[*]		Appx:SA1[±]	
In	Out[†]	Out[‡]	$e^{\ddagger}$	Out[*]	$e^{\star}$	Out[±]	$e^{\pm}$
00000	000	000	0	000	0	001	1
00001	001	001	0	000	1	001	0
00010	010	010	0	010	0	011	1
00011	011	011	0	010	1	011	0
00100	001	001	0	000	1	001	0
00101	010	000	2	000	2	001	1
00110	011	011	0	010	1	011	0
00111	100	010	2	010	2	011	1
01000	010	010	0	010	0	011	1
01001	011	011	0	010	1	011	0
01010	100	100	0	100	0	101	1
01011	101	101	0	100	1	101	0
01100	011	011	0	010	1	011	0
01101	100	010	2	010	2	011	1
01110	101	101	0	100	1	101	0
01111	110	100	2	100	2	101	1
10000	001	001	0	000	1	001	0
10001	010	000	2	000	2	001	1
10010	011	011	0	010	1	011	0
10011	100	010	2	010	2	011	1
10100	010	000	2	000	2	001	1
10101	011	001	2	000	3	001	2
10110	100	010	2	010	2	011	1
10111	101	011	2	010	3	011	2
11000	011	011	0	010	1	011	0
11001	100	010	2	010	2	011	1
11010	101	101	0	100	1	101	0
11011	110	100	2	100	2	101	1
11100	100	010	2	010	2	011	1
11101	101	011	2	010	2	011	2
11110	110	100	2	100	2	101	1
11111	111	101	2	100	3	101	2

[†]Golden non-approximated 2-bit adder response
[‡]Approximated adder response (carry cut)
[*]Approx adder with SA0 fault at sum_0 (f1$_{SA0}$)
[±]Approx adder with SA1 fault at sum_0 (f1$_{SA1}$)
Input bits : $\{C_{in}a_1a_0b_1b_0\}$, Output bits : $\{C_{out}sum_1sum_0\}$
e: error in each case, worst-case errors $e_{wc}^{\ddagger} = 2$, $e_{wc}^{\star} = 3$, $e_{wc}^{\pm} = 2$

structure is built. The procedure $get\_network()$ does this part and the data structure is a DAG preserving the individual gate details. Further, the algorithm iterates through each fault in the input fault list. The procedure $get\_faulty\_network()$

Algorithm 6.1 Approximation-aware fault classification

```
 1: function APPROX_PREPROCESS(faultList faults, error behavior e)
 2:     C ← get_network()
 3:     for each f ∈ faults do
 4:         if fault_not_processed(f) then
 5:             Ĉ ← get_faulty_network(f)
 6:             E ← get_error_computation_network(metric(e))
 7:             D ← negation_of(e)
 8:             φ = construct_miter(C, Ĉ, E, D)
 9:             result = solve(φ)
10:             if result = SAT then
11:                 set f_status ← NonApproxFault
12:             else
13:                 set f_status ← ApproxFault
14:             end if
15:             imply_approximation (f, f_status)
16:         end if
17:     end for
18:     return faults
19: end function
```
---

takes in this fault and modifies the netlist based on the fault type. For a SA0 fault the signal line (the wire corresponding to that signal) is tied to logic-0 and SA1 to logic-1. The *get_error_computation_network*() encodes the input error metric information to an error computation network. Further, the procedure *negation_of*() negates the output of this encoding. This is same as appending an inverter at the output signal. The error computation network is specific to the type of error metric under consideration such as worst-case error or bit-flip error. For detailed explanation on the internals of the error computation refer Chap. 3. The purpose and specific use of these procedures in the context of testing will be more clear once the details on the *approximation-aware miter* are provided in the explanation to follow.

The core of the algorithm is to construct an approximation-aware miter for fault classification (see Line 8). This formulation is then transformed into a SAT instance which is solved by a SAT solver. This SAT instance is the Conjunctive Normal Form that a SAT solver works with. The general principle of an approximation miter has already been presented in Chap. 3 where the error metrics are precisely computed. In this work, however, the miter principle is followed but used to determine the fault classification. After fault classification, structural techniques are applied to deduce further faults. The pre-processor algorithm returns the same list of faults, but for each fault the status has been updated, i.e., it has been classified as approximation-redundant or non-approximation. In the following, we explain how the approximation miter for fault classification is constructed and used in our approach. The general form of the approximation miter is reproduced in Fig. 6.3 in the context of testing to facilitate better understanding of individual steps involved.

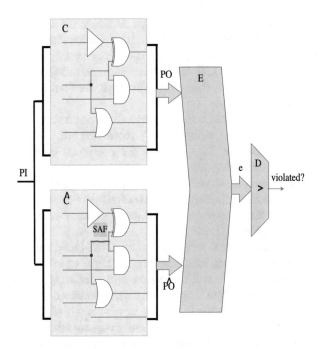

Fig. 6.3 Approximation Miter for Approximation-aware Fault Classification

6.2.2.2 Approximation Miter for Fault Classification

The approximation miter for fault classification (see Fig. 6.3 and Line 8 in Algorithm 6.1) is constructed using

- the *golden reference netlist* C—consists of the correct (non-approximated) circuit (provided by get_network() in Line 2)
- the *faulty approximate netlist* $\hat{C}$—this netlist is the final approximate netlist including fault f (provided by get_faulty_network(f) in Line 5)
- the *error computation network* E—based on the given error metric, this network is used to compute the concrete error of a given output assignment of both netlists (see Line 6)
- the *decision network* D—the result of the decision network becomes 1, if the comparison of both netlists violates the error metric constraint

Again, the goal of the proposed miter for approximation-aware fault classification is to decide whether the current fault is approximation-redundant or not. In other words we are looking for an input assignment such that the given error metric constraint is violated. For instance, in case of the motivating example (Fig. 6.2) this worst-case constraint is $e_{wc} \leq 2$, so we are looking for its negation. For this approximate adder example we set D to $e_{wc} > 2$ (see again Line 7).

Now the complete problem is encoded as a SAT instance and run a SAT solver. If the solver returns satisfiable—so there is at least one input assignment for which the result violates the error metric constraint—it is proven that the fault is a non-approximation fault (Line 11). If the solver returns unsatisfiable, the fault is an approximation-redundant fault (Line 13). This fault does not have to be targeted during the regular ATPG stage.

In addition to the SAT techniques mentioned above, several structural techniques are also used in conjunction with the SAT solver for efficiency (see Line 15). This includes, for example, fault equivalence rules and constant propagation for redundancy removal. Besides, several trivial approximation-redundant/non-approximation faults can be identified. Such trivial faults are located near the outputs. An example is a fault affecting the MSB output bits that always results in error metric constraint violation. These can be directly deduced as non-approximation faults through path tracing.

It is important to point out the significance of a SAT-based methodology for the proposed approximation-aware test. Another technique which is most commonly employed in an ATPG tool is fault simulation. However, fault simulation by itself cannot guarantee whether a fault is approximation redundant or not. The simulation has to be continued exhaustively for all the combinations of input patterns until a violation of the error metric is observed. For an approximation-redundant fault, this will end up simulating all the input patterns invariably, i.e., 2^n combinations for a circuit with input n-bits. Clearly this is infeasible and impractical. Similarly, it is common in classical ATPG approaches to employ fault simulation on an initially generated test set to detect further faults. The tools read in an initial set of ATPG patterns and do a fault simulation to detect a subset of the remaining faults. However, this approach also cannot be used for approximation-aware fault classification since the fault manifestation, i.e., the propagation path, is only one among many possibilities. Hence, in this case, individual faults have to be targeted for classification one at a time.

In the next section the experimental results are provided.

6.3 Experimental Results

All the algorithms are implemented as part of the *aXc* framework.[1] The input to our program is the gate level netlist of the approximated circuit which is normally used for standard ATPG generation. Now, instead of running ATPG, the approximation-aware fault classification approach (cf. Sect. 6.2) is executed. This filters out the approximation-redundant faults. From there on the standard ATPG flow is taken.

---

[1]Note: The approximation-aware test feature for the *aXc* tool is proprietary and not publicly available.

In the following, the results for approximated circuits using worst-case error and bit-flip error constraints are provided. The experiments have been carried out on an Intel Xeon CPU with 3.4 GHz and 32 GB memory running Linux 4.3.4. Considering error-rate is left out for future work. As explained in the previous chapters, the error-rate depends on *model counting*. Model counting is a higher complexity problem compared to SAT (#P-complete vs NP-complete). Hence, it is computationally intractable to invoke a model counting engine for each of the faults in the fault-list due to the huge volume of the faults. We refer to Sect. 3.4 in Chap. 3 for further details on computational complexity of the individual error metrics.

The experimental evaluation of our approach has been done for a wide range of circuits. For the circuits the respective error metrics are obtained from the *aXc* tool using the techniques explained in the earlier chapters. In this section, first the results using the worst-case error as approximation pre-processing criteria are explained. These results are provided in Table 6.2. The experimental evaluation using the bit-flip error metric is separately explained at the end of this section in Table 6.4. Note that a combination of these error metrics can also be provided to the tool. Further the worst-case error and the bit-flip error are the error metrics coming from the application.

6.3.1   Results for the Worst-Case Error Metric

All the results for the worst-case error scenario are summarized in Tables 6.2 and 6.3. The general structure of these tables is as follows. The first three columns give the circuit details such as the number of primary inputs/outputs and the gate count. The gate count is from the final netlist to which ATPG is targeted. This is followed by the fault count without the approximation-aware fault classification, i.e., this gives the "normal" number of faults for which ATPG is executed. The regular fault-equivalence and fault-dominance are already accounted in these fault counts (column: f_{orig}). The next two columns provide the resulting fault count and reduction in faults using the approximation-aware fault classification methodology (columns: f_{final}^{wc} and $f_{\Delta}^{wc}\%$). The last column denotes the run time in CPU seconds spent for the developed approach, i.e., only the pre-processing (Algorithm 6.1). Altogether, there are four different sets of publicly available benchmark circuits where the approximation-aware test is evaluated.

6.3.1.1   Arithmetic Circuits

Table 6.2 consists of commonly used approximation arithmetic circuits. The first set is manually architected approximation adders such as *Almost Correct Adder* (ACA adder) and *Gracefully Degrading adder* (GDA adder). The authors of these works have primarily used these adders in image processing applications [KK12, ZGY09, YWY+13]. These designs are available in the repository [SAHH15]. A summary of the error characteristics of these adders is already provided in Sect. 3.5.1.1 in Chap. 3.

Table 6.2 Summary of approximation-aware fault classification for worst-case error: benchmarks set-1

Benchmark details			#Faults			Time
Circuit	#PI/#PO	#Gates	f_{orig}	f_{final}^{wc} [†]	$f_{\Delta}^{wc}(\%)$	s
Architecturally approximated adders[1] (set:1)						
ACA_II_N16_Q4 [±]	32/17	225	483	180	62.73	14
ACA_II_N16_Q8	32/17	255	535	277	48.22	16
ACA_I_N16_Q4	32/17	256	530	174	67.17	14
ETAII_N16_Q8 [∓]	32/17	255	535	277	48.22	16
ETAII_N16_Q4	32/17	225	483	180	62.73	13
GDA_St_N16_M4_P4[‡]	32/17	258	575	331	42.43	17
GDA_St_N16_M4_P8	32/17	280	617	188	69.53	21
GeAr_N16_R2_P4[‡‡]	32/17	255	541	160	70.43	16
GeAr_N16_R6_P4	32/17	263	561	286	49.02	19
GeAr_N16_R4_P8	32/17	261	552	161	70.83	17
GeAr_N16_R4_P4	32/17	255	535	277	48.22	16
Arithmetic designs[2] (set:2)						
Han Carlson Adder*	64/33	655	1415	969	31.52	88
Kogge Stone Adder*	64/33	839	1789	1475	17.55	140
Brent Kung Adder*	64/33	545	1178	700	40.58	51
Wallace Multiplier*	16/16	641	1641	669	59.23	5027
Array Multiplier*	16/16	610	1585	619	60.95	4250
Dadda Multiplier*	16/16	641	1641	652	59.40	6875
MAC unit1*	24/16	725	1821	760	58.26	12,782
MAC unit2*	33/48	874	2104	492	76.61	921
4-Operand Adder*	64/18	614	1434	1156	19.39	60

#PI, #PO: number of primary inputs, outputs. #gates: gate count after synthesis
f_{orig}: final fault count for which ATPG generated without approximation
(dominant, equivalent faults not included)
f_{final}^{wc} : final fault count after approximation pre-processor with worst-case error limits

f_{Δ}^{wc}: relative reduction in the fault count in %. $f_{\Delta}^{wc} = \left(\dfrac{f_{orig} - f_{final}^{wc}}{f_{orig}} \right) * 100$

time: time taken for f_{final}^{wc}, [†] worst-case error evaluated using *aXc* SAT techniques
*shows automated approximation synthesis technique using *aXc*
[1] Ad-hoc architecturally approximated adders: [±]ACA adder [KK12], [∓]ETA adder [ZGY09]
[‡]GDA adder [YWY+13], [‡‡]GeAr adder [SAHH15], set:2 arithmetic benchmarks from [Aok16]

As evident from Table 6.2, a significant portion of the faults in all these designs are approximation-redundant. It can also be seen that such architectural schemes show a wide range in approximation-redundant fault count, even in the same category. For example, among the different *Almost Correct Adders* [KK12], ACA_I_N16_Q4 has a far higher ratio of approximation faults compared to the scheme ACA_II_N16_Q8 (67% vs 48%). The adder GDA_St_N16_M4_P4 [YWY+13] has the least ratio of approximation faults in this category, about 42%.

Table 6.3 Summary of approximation-aware fault classification for worst-case error: benchmarks set-2

Benchmark details			#Faults			Time
Circuit	#PI/#PO	#Gates	f_{orig}	f_{final}^{wc} †	$f_{\Delta}^{wc}(\%)$	s
EPFL benchmarks[3] (set:3)						
Barrel shifter*	135/128	3975	8540	6677	21.81	3493
Max*	512/130	3780	7468	5783	22.56	2156
Alu control unit*	7/26	178	378	252	33.33	5
Coding-cavlc*	10/11	885	1830	1194	34.75	73
Lookahead XY router*	60/30	370	739	459	62.11	12
Adder*	256/129	1644	3910	2738	29.97	969
Priority encoder*	128/8	1225	2759	1335	51.61	84
Decoder*	8/256	571	2338	2175	6.97	132
Round robin*	256/129	16,587	26,249	11,802	55.04	43,940
Sin*	24/25	5492	13,979	12,756	8.74	7464
Int to float converter*	11/7	296	624	464	25.64	7
ISCAS-85 benchmarks[4] (set:4)						
c499*	41/32	577	1320	755	42.80	53
c880*	60/26	527	1074	271	74.77	27
c432*	36/7	256	487	441	09.45	7
c1355*	41/32	575	1330	680	48.87	57
c1908*	33/25	427	974	694	28.74	46
c2670*	233/140	931	1950	372	80.92	138
c3540*	50/22	1192	2657	2388	10.12	268
c5315*	178/123	2063	4224	2851	32.50	1112
c7552*	207/108	2013	4490	2938	34.57	1014

#PI, #PO: number of primary inputs, outputs. #gates: gate count after synthesis
f_{orig} : final fault count for which ATPG generated without approximation
(dominant, equivalent faults not included)
f_{final}^{wc} : final fault count after approximation pre-processor with worst-case error limits

f_{Δ}^{wc}: relative reduction in the fault count in %. $f_{\Delta}^{wc} = \left(\dfrac{f_{orig} - f_{final}^{wc}}{f_{orig}} \right) * 100$

time: time taken for f_{final}^{wc}, †worst-case error evaluated using *aXc* SAT techniques
∗ shows automated approximation synthesis technique using *aXc*

In the second set, other arithmetic circuits such as fast adders, multipliers, and multiply accumulate (MAC) are evaluated. These designs are taken from [Aok16]. The approximate synthesis techniques provided in the previous chapter are used to approximate these circuits. Similar to the architecturally approximated designs, the relative mix of approximation-redundant and non-approximation faults in these circuits also vary widely depending on the circuit structure.

6.3.1.2   Other Standard Benchmark Circuits

The approximation-aware fault classification is also evaluated on circuits from the ISCAS-85 [HYH99] and EPFL [AGM15] benchmarks to demonstrate its generality. These results are provided as set:3 and set:4 in Table 6.3. A high percentage of faults is classified as approximation-redundant and these faults can be skipped from ATPG generation, eventually improving the yield. The highest fraction of approximation-redundant faults is obtained in the iscas-85 circuit c2670 (above 80%). However, there is a wide variation in the relative percentage of faults classified as approximation-redundant. This primarily stems from the structure of the circuit, approximation scheme employed, and the error tolerance of the end application.

6.3.2   Results for the Bit-Flip Error Metric

The bit-flip error is another important approximation error metric. The bit-flip error is independent of the error magnitude and relates to the hamming distance between the golden non-approximated output and the approximated one. The same set of designs given in Tables 6.2 and 6.3 are used to evaluate the approximation-aware fault classification methodology under the bit-flip error metric. The results obtained are summarized in Table 6.4. Table 6.4 shows the approximation-aware fault classification results for architecturally approximated adders [KK12, ZGY09, YWY+13, SAHH15], arithmetic designs [Aok16], standard ISCAS benchmark circuits [HYH99], and EPFL benchmarks [AGM15].

The results in Table 6.4 show a different trend compared to the worst-case error results in Table 6.2. As mentioned before the bit-flip error is the maximum hamming distance of the output bits of the approximated and non-approximated designs, irrespective of the error magnitude. In general, the approximation pre-processor has classified a lesser percentage of faults as approximation redundant in the first category of hand-crafted approximated adder designs. This has to be expected since each approximation scheme is targeted for a different error criteria, and therefore has a different sensitivity for each of these error metrics. Furthermore, these two error metrics are not correlated. As an example, a defect affecting only the most significant output bit has the same bit-flip error as that of a defect affecting the least significant output bit of the circuit. However, the worst-case errors for these respective defects are vastly different. The individual works [KK12, ZGY09, YWY+13, SAHH15], etc. can be referred for a detailed discussion of the error criteria employed in the design of these circuits. Nevertheless, the approximation-aware fault classification tool has classified many errors as approximation-redundant for several circuits as can be seen in Table 6.4.

Overall, the results confirm the applicability of the proposed methodology. The methodology can be easily integrated into today's standard test generation flow. Note that, in general the run times for a SAT-based ATPG flow depend mainly on

Table 6.4 Summary of the approximation-aware fault classification results for the bit-flip error

Benchmark	#Faults			Time
Adders[1]	f_{orig}	f_{final}^{bf}	f_{Δ}^{bf} (%)	s
ACA_II_N16_Q4	483	400	17.18	4
ACA_II_N16_Q8	535	480	10.28	4
ACA_I_N16_Q4	530	426	19.62	5
ETAII_N16_Q8	535	480	10.28	5
ETAII_N16_Q4	483	400	17.18	4
GDA_St_N16_M4_P4	575	508	11.65	5
GDA_St_N16_M4_P8	617	197	68.07	7
GeAr_N16_R2_P4	541	528	2.40	5
GeAr_N16_R6_P4	561	200	64.35	5
GeAr_N16_R4_P8	552	199	63.95	6
GeAr_N16_R4_P4	535	480	10.28	5
EPFL circuits[3]	f_{orig}	f_{final}^{bf}	f_{Δ}^{bf} (%)	s
Barrel shifter*	8540	3454	59.55	61,488
Alu control unit*	378	178	52.91	11
Coding-cavlc*	1830	1346	26.45	76
Lookahead XY router*	739	655	11.37	77
Int to float converter*	624	293	53.04	9
Priority encoder*	2759	1061	61.54	87
Round robin	26,249	11,802	55.04	43,940

Benchmark	#Faults			Time
Arith designs[1]	f_{orig}	f_{final}^{bf}	f_{Δ}^{bf} (%)	s
Han Carlson Adder*	1415	1202	15.05	155
Kogge Stone Adder*	1789	1699	5.03	105
Brent Kung Adder*	1178	1018	13.58	58
Wallace Multiplier*	1641	309	81.17	52
Array Multiplier*	1585	311	80.37	55
Dadda Multiplier*	1641	303	81.13	54
MAC unit1*	1821	1775	2.53	70
MAC unit2*	2104	2017	4.13	161
4-Operand Adder*	1434	1332	7.11	47
ISCAS-85[3]	f_{orig}	f_{final}^{bf}	f_{Δ}^{bf} (%)	s
c499*	1320	1153	12.65	73
c880*	1074	305	71.60	31
c432*	487	480	1.44	3
c1355*	1330	1196	10.08	79
c1908*	974	949	2.57	30
c2670*	1950	428	78.05	396
c3540*	2657	839	68.42	418

Note: Other details on the circuits are available in Tables 6.2 and 6.3

f_{orig}: original fault count for ATPG without bit-flip approximation

(Note: dominant and equivalent faults are excluded from this count)

f_{final}^{bf}: final fault count after approximation-aware fault classification

f_{Δ}^{bf} (%): Reduction in fault count $= \left(\dfrac{f_{orig} - f_{final}^{bf}}{f_{orig}} \right) * 100$

time: processing time taken by approximation-aware fault classification

* shows automated approximation synthesis technique using aXc

the circuit complexity, size, and the underlying SAT techniques [BDES14]. The approximation-aware test approach is also influenced by these factors. Therefore, improvements in SAT-based ATPG have a direct impact in our approach. To this end, several advanced techniques have been proposed. A detailed overview of such techniques can be found in [SE12]. It is also worth mentioning that the approximation-aware fault classification and the subsequent ATPG generation is a one time effort whereas the actual post-production test of the circuit is a recurring one. Hence, the additional effort and run times are easily justified due to high reduction in the fault count that has to be targeted for test generation.

6.4 Concluding Remarks

In this chapter, we presented an approximation-aware test methodology. To the best of our knowledge, this is the first work that examines the impact of design level approximations in post-production test. First, we proposed a novel fault classification based on the approximation error characteristics. Further, we showed a formal methodology that can map all the faults in an approximation circuit into approximation-redundant and non-approximation faults. The approximation-redundant faults are guaranteed to have effects that are below the error threshold limits of the application. Hence, the subsequent ATPG generation has to target only the non-approximation faults and thereby yield can be improved significantly. Our methodology can be easily integrated into today's standard test generation flow. Besides, the experimental results on a wide range of circuits confirm the potential and significance of our approach. Substantial reduction in fault count up to 80% is obtained depending on the concrete approximation and the error metric.

Chapter 7
ProACt: Hardware Architecture for Cross-Layer Approximate Computing

The previous chapters in this book focused on CAD tools for approximate computing. These included the algorithms and methodologies for verification, synthesis, and test of an approximate computing circuit. However, the scope of approximate computing is not limited to such circuits and the hardware built from them. In fact, approximate computing spans over multiple layers from architecture and hardware to software. There are several system architectures proposed ranging from those employing neural networks to dedicated approximation processors [YPS$^+$15, SLJ$^+$13, VCC$^+$13, CWK$^+$15]. The hardware is designed wrt. an architectural specification on the respective error criteria. Further, the software that runs the end application is aware of such hardware features and actively utilizes them. This is called *cross-layer approximate computing*, where the software and hardware work in tandem according to an architectural specification [VCC$^+$13]. Such systems can harness the power of approximations more effectively [VCRR15, ESCB12]. This chapter details an important microprocessor architecture developed for approximate computing. This architecture ProACt can do cross-layer approximations spanning hardware and software. ProACt stands for *Processor for On-demand Approximate Computing*. Details on this processor architecture, implementation, and the detailed evaluation are provided in the forthcoming sections.

7.1 Overview

Approximate computing can deliver significant performance benefits over conventional computing by relaxing the precision of results. As mentioned before, in order to harness the full potential of approximations, both hardware and software need to work in tandem [XMK16, ESCB12]. However, in a general application, there are program segments that can be approximated and others which are critical and should not be approximated. In addition, it has become a good strategy to perform

© Springer Nature Switzerland AG 2019
A. Chandrasekharan et al., *Design Automation Techniques for Approximation Circuits*, https://doi.org/10.1007/978-3-319-98965-5_7

approximation in certain situations, e.g., when the battery goes low. Also, decisions such as the duration and degree of approximations may depend on external factors and input data set of the application, which may not be fully known at the system design time. This can also be driven by the user of the end application. All these call for an *on-demand*, rapidly switchable hardware approximation scheme that can be fully controlled by software. This software control may originate from the application itself or from a supervisory software like the operating system. Software techniques and methodologies for cross-layer approximate computing have been extensively studied in the literature [HSC+11, BC10, CMR13, SLFP16]. These software methodologies can utilize the underlying hardware approximations more effectively than conventional approaches.

The ProACt processor architecture fulfills all the above requirements—hardware approximations only when needed with software control over the degree and extend of approximations. ProACt stands for *Processor for On-demand Approximate Computing*. The core idea of ProACt is to functionally approximate floating point operations using previously computed results from a cache thereby relaxing the requirement of having the exact identical input values for the current floating point operation. To enable on-demand approximation a *custom instruction* is added to the *Instruction Set Architecture* (ISA) of the used processor which essentially adjusts the input data for cache look-up and by this controls the approximation behavior. The approach and achieved results have been published in [CGD17a].

ProACt is based on the RISC-V instruction set architecture [WLPA11]. RISC-V is a state-of-the-art open-source 64 bit RISC architecture. The current hardware implementation of ProACt is a 64-bit processor with 5 pipeline stages. ProACt has integrated L1 and L2 cache memories, a dedicated memory management unit (MMU), and DMA for high bandwidth memory access. Overall, a ProACt application development framework is created for developing the software targeted to ProACt. The framework, shown in Fig. 7.1, consists of the extended hardware processor and the software tool chain to build a complete system for on-demand approximate computing. The software tool chain is based on GNU Compiler Collection (GCC) and includes the standard development tools such as compiler, linker, profiler, and debugger.[1] Besides a set of approximation library routines are also provided as an Application Program Interface (API) for the easier development of the applications. For quicker turn-around times, a ProACt emulator is also available which does a cycle-accurate emulation of the underlying ProACt hardware.

In ProACt floating point operations are primarily targeted for approximation. There are several different standards for the hardware representation of a real number such as fixed-point arithmetic and floating-point arithmetic. Fixed point arithmetic has historically been cheaper in hardware, but comes with less precision and lower dynamic range. The current implementation of ProACt supports only floating point arithmetic. Many state-of-the-art signal processing applications

---

[1]Currently only the development tools for C/C++ and assembly languages are developed. The ProACt compiler tool set is based on the GCC version 6.1.

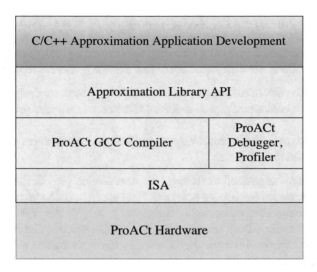

Fig. 7.1 ProACt application development framework

(e.g., speech recognition, image rendering, etc.) are complex, and these demand a wide dynamic range of signals only available with floating point systems. As a result, many commercial signal processing engines incorporate floating point units [Oma16] and even the industry standards are written mainly for floating point arithmetic [Itu16]. However, complex floating point operations such as division and square-root are computationally expensive, and usually span over multiple clock cycles [Mos89, CFR98, Hau96].

The main aim of ProACt is to reduce the number of clock cycles spent on floating point operations with the on-demand approximation scheme. Therefore, these operations and results are stored in an approximation look-up table. This look-up table is checked first before executing a new floating point operation. In this step, *approximation masks* are applied to the operands before the cache look-up. These masks are set by the new custom approximation control instruction (assembly mnemonic SXL[2]), and they define the degree of approximation. The SXL instruction also controls additional flags for approximation behavior such as enabling/disabling the cache look-up table. Thus, ProACt functions as a normal processor when the approximation look-up table is disabled.

The custom approximation control instruction SXL is designed as an *immediate* instruction resulting in very little software overhead and run-time complexity. SXL is fully resolved at the *decode* stage of the processor pipeline, and therefore situations such as *control* and *data hazards* [PH08] are reduced to a minimum. This is significant in a multi-threaded, multi-process execution environment, and in atomic operations. Being able to rapidly switch between approximation and normal modes is an important factor affecting the throughput of the processor in such contexts.

---

[2]SXL stands for *Set approXimation Level.*

A hardware prototype of ProACt is implemented in a Xilinx FPGA. Experimental results show the benefits in terms of speed for different applications. The complete ProACt development framework, including the hardware prototype details and the sample applications, are distributed open source in the following repositories:

- https://gitlab.com/arunc/proact-processor : ProACt processor design
- https://gitlab.com/arunc/proact-zedboard : Reference hardware prototype implementation
- https://gitlab.com/arunc/proact-apps : Application development using ProACt and approximation libraries.

This chapter is organized as follows: the next section provides an overview of the existing literature on floating point units and approximation processors. Cross-layer approximate computing architectures are a very active research area. We summarize the major advancements in this field in this section. The ProACt system architecture is explained afterwards, followed by the experimental evaluation on an FPGA prototype.

7.1.1 *Literature Review on Approximation Architectures*

There are several works in approximate computing for hardware and software approximations ([ESCB12, VCC+13, CWK+15, YPS+15] etc.), that range from dual voltage hardware units to dedicated programming languages. In particular [ESCB12] explains a generalized set of architectural requirements for an approximate ISA with a dedicated programming language. The authors discuss approximation variants of individual instructions that can be targeted to a dual voltage hardware showing the benefits, though only in simulation. Similarly, significant progress has been made on custom designed approximation processors such as [SLJ+13, CDA+12, KSS+14] targeted for one particular area of application. In contrast, ProACt is a general purpose processor with on-demand approximation capabilities. Approximations can even be completely turned off in ProACt. Moreover, our work is focused on functional approximations, rather than schemes like dynamic voltage/frequency scaling (DVFS), that involve fabrication aspects, and are tedious to design, implement, Test, and characterize. A detailed overview of such techniques is available in [XMK16, VCRR15].

Several schemes have been presented so far to approximate a floating point unit. Hardware look-up tables are used in [CFR98] to accelerate multi-media processing. This technique called *memoing* has been shown to very effective, though it does not use any approximation. The work of [ACV05] extends this further to *fuzzy memoization*, where approximations are introduced in the look-up table, thereby increasing the efficiency greatly. However, none of these approaches use custom instructions for software controlled approximations. Therefore, the scope of such systems is limited only to the application it is designed for, like multi-media data processing. In addition, these approaches do not offer direct support to treat the critical program segments differently or to limit approximations only to non-critical sections of the program. An example from image processing is the JPEG image header which contains critical information, and should not be approximated, whereas the pixel data content of the image is relatively safe to approximate.

There have been approaches in the domain of *Application Specific Instruction Set Processors* (ASIP) for approximate computing. For instance, [KGAKP14] uses custom instructions for approximations that map to dedicated approximation hardware. Several custom approximation hardware units are presented and the goal is to select custom instructions and hardware which conform to a predefined quality metric. Altogether a different approach to approximations is used in ProACt, where only a few light weight custom instructions are provided which enable, disable, or control the level of precision of approximations.

ProACt has several advantages. A supervisory software such as an operating system can solely decide on the approximation accuracy and the applications need not be aware of this. Thus, the same applications compiled for a standard ISA can be run without approximations, with approximations, and even to a varying level of precision. Furthermore, even if the application binary is generated with custom approximation-control instructions, the overhead incurred is very little compared to an ASIP implementation. ProACt can respond and adapt to varying conditions with very less software complexity. A simple *if statement* is usually only required in such situations to bring in the power of approximations, as compared to a whole program with special custom instructions.

A detailed overview on ProACt system architecture is given in the next section.

7.2 ProACt System Architecture

In this section, the important architectural features of ProACt are described. The ProACt system overview is shown in Fig. 7.2 on the left-hand side. As can be seen it consists of the processor hardware and the software units working together to achieve approximation in computations. To operate the approximations in hardware an *Approximate Floating Point Unit* (AFPU) is added to the architecture. A zoom is given on the right-hand side of Fig. 7.2 to show this AFPU. Its details are described in the next section. In the normal mode (i.e., approximations disabled), ProACt floating point results are IEEE 754 compliant. In this scheme, a double precision format, `double`, is 64-bit long with the MSB sign-bit followed by a 11-bit exponent part, and a 52-bit fraction part. In the following discussion, all the numbers are taken to be `double`, though everything presented applies to other floating point representations as well.

The AFPU is explained next, followed by the ISA extensions for approximation. The ISA and the assembly instructions are the interface between ProACt hardware and the applications targeting ProACt. Other details on the processor architecture and compiler framework are deferred to the end of this section.

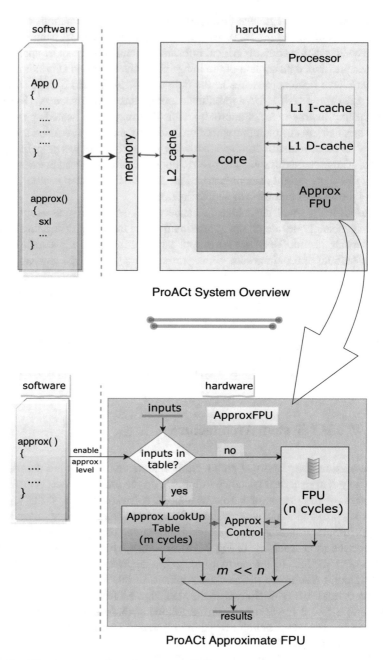

Fig. 7.2 ProACt system overview: software controlled hardware approximations

7.2.1 Approximate Floating Point Unit (AFPU)

The AFPU in ProACt consists of an approximation look-up table, a pipelined *Floating Point Unit* (FPU), and an approximation control logic (see Fig. 7.2, right-hand side). The central approach used in this work is that the results of the FPU are stored first, and further operations are checked in this look-up table, before invoking the FPU for computation. The input arguments to the FPU are checked in the look-up table and when a match is found, the results from the table are fed to the output, bypassing the entire FPU. The FPU will process only those operations which do not have results in the table. This look-up mechanism is much faster, resulting in significant savings in clock cycles. Approximation masks are applied to the operands before checking the look-up table. Thus, the accuracy of the results can be traded-off using these masks. These approximation masks are set by the software (via the custom approximation control instruction SXL, for details see next section) and vary in precision. The mask value denotes the number of bits to be masked before checking for an entry in the look-up table. The mask value is alternatively called *approximation level*. The number of bits masked is counted from the LSB of the operands. For example, if the approximation level is 20, the lower 20-bits of the fraction part will be masked before querying the look-up table. The figure below shows a standard IEEE double standard representation and the same representation with 20-bits masked as approximation.

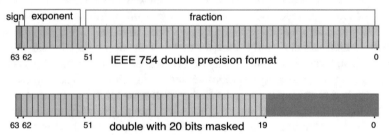

These approximation levels are fully configurable and provide a bit level granularity to the approximations introduced. Software controls the approximation mask, look-up table mechanism, and all these units can be optionally turned off.

ProACt uses an in-order pipelined (see Sect. 7.2.3 for hardware details) architecture to improve performance. Hence, the results from the AFPU have to be in the order of the input operations supplied. This in-order execution of the AFPU is ensured by the approximation control unit. While in action, some operations will be dispatched to the FPU whereas some others will be resolved to the approximation look-up table, subsequently achieving lesser cycles for results. Note that the operations resolved to the approximation look-up table are the cached results of earlier FPU operations. The final results are ordered back into the input order by the approximation control logic.

The look-up table stores the last N floating point operations in a round robin fashion. Several real-world data has high degree of *spatial* and *temporal locality*.

For example, the chances that the neighboring pixels in an image have similar content are high (spatial locality). In many algorithms this directly translates to a corresponding temporal locality since the pixels are processed in a defined order and not randomly. This is also exploited in the ProACt scheme, where the $(N + 1)$th result simply overwrites the first result, preserving the last N results always.[3]

As mentioned before, the software controls the hardware approximation mechanism in ProACt. This is achieved by extending the ISA with a custom instruction for approximation. This ISA extension is presented in the next section.

7.2.2   Instruction Set Architecture (ISA) Extension

The software compiler relies on the ISA to transform a program to a binary executable. Hence, the ISA is extended with a single assembly instruction SXL (*Set approXimation Level*) for the software control of approximations. SXL is designed as an immediate instruction that takes a 11-bit immediate value. The LSB, when set to "1", enables the hardware approximations. The remaining bits are used to set the approximation level and other special flags.

ProACt floating point operations can come under three categories. First one is the normal approximation operation where the mask value is a non-zero number and the look-up table is enabled. ProACt can also operate with a 0-mask value and look-up table enabled. This mode results in exact computations similar to approximations disabled, since the operands and thereby the results from the previous computations have to match exactly in the look-up table. It is worth noting that this non-approximating cache look-up mode of ProACt can also potentially speed up computations for several applications as shown in [CFR98]. The approximation masks applied to the input operands further improvise on this as the look-up table hit-rate and the computation reuse increases with the degree of approximation. The third mode is the look-up table and approximations completely disabled, whereupon ProACt works like a normal processor used in the conventional computing.

SXL is fully resolved in the decode stage of the pipeline in the hardware micro-architecture level. Since the instruction is designed as an immediate instruction, there are no side effects such as memory operations and register read/write that needs to be taken care of in later stages of the pipeline such as execute and write-back. Similarly, data hazards and control hazards due to SXL are minimal since there are no other dependencies. This simplifies the processor design as a whole and improves the pipelined throughput of the processor. Besides, the instruction itself is very light weight and the processor can easily enable/disable approximation. This is important for atomic operations and also helps rapid context switching for critical program segments.

---

[3]In future, look-up table update policy will be configurable through SXL and schemes like LRU (*Least Recently Used*) will be supported.

7.2.3   ProACt Processor Architecture

The ProACt processor is based on the RISC-V architecture [WLPA11]. RISC-V is a modern, general purpose, high quality instruction set architecture based on RISC principles. The ISA under the governance of the RISC-V Foundation is intended to become an industry standard.[4] Further, RISC-V is distributed open source, thus making it well suited for academic and research work. The RISC-V supports several extensions for both general purpose and special purpose computing. Out of these, ProACt supports integer multiplication and division, atomic instructions for handling real-time concurrency, and IEEE floating point with double precision. All the memory operations in ProACt are carried out through load/store instructions. Further, all the memory accesses are *little-endian*, i.e., the least significant byte has the smallest address.

Several implementations of the RISC-V ISA are publicly available. ProACt is based on one such implementation called Rocket chip [A⁺16], which is described in the *Chisel* hardware description language [BHR⁺12]. The acronym "Chisel" stands for *Constructing Hardware in a Scala Embedded Language*. As the name indicates, Chisel is essentially a *Domain Specific Language* (DSL) built on top of the Scala programming language. Chisel has several advanced features to support hardware development such as functional programming, object orientation, and parametrized generators. ProACt is also developed in Chisel and inherits several features from the Rocket chip SoC. The high level design in Chisel is converted to a synthesizable Verilog RTL description with the help of a Chisel compiler.

ProACt uses 64-bit addressing scheme with 32 general purpose registers.[5] ProACt has 32 dedicated floating point registers. In addition to this, a set of control registers are also available. Two important control registers are *mcycle* and *minstret*. These registers can be used for tracking the hardware performance. For further details, we refer to the ProACt documentation and the RISC-V manual [WLPA11]. The pipeline used is a 64-bit, 5-stage, in-order pipeline. The pipeline design largely follows a classic RISC pipeline with stages being Instruction Fetch (IF), Instruction Decode (ID), Execute (EX), Memory access (MEM), and Write Back (WB) [HP11]. As shown in Fig. 7.2, the processor has separate L1 caches for instruction and data, and a unified L2 cache memory. The *Memory Management Unit* (MMU) supports page-based virtual memory addressing and DMA for high bandwidth memory access.

---

[4]http://riscv.org.

[5]Note: Register zero is the constant 0. By design, all reads from this register will result in a value "0" and all writes are discarded. This is a widely adopted practice in RISC CPU design.

7.2.4   Compiler Framework and System Libraries

The ProACt compiler framework consists of a cross-compiler, a linker, and associated tools based on *GNU Compiler Collection* (GCC) [Gcc16] and GNU Binutils.[6] Further, the *newlib* library[7] is opted for the current ProACt compilers primarily targeting embedded system developers. To build the ProACt GCC cross-compiler from sources, a standard C++ compiler is required. The version of the tool set currently developed for ProACt corresponds to the 6.1 version of GCC. The ProACt cross-compiler is used like a regular compiler. This means that, currently there are no special options for approximations to be passed on to the compiler. Rather, the on-demand approximation scheme is implemented using a set of library routines, which the user calls from the application.

As mentioned before, a set of system library routines and macros are provided with the distribution for the convenient use of on-demand approximations in software programming. These routines can be used to enable/disable the on-demand approximation feature, control the approximation look-up table mechanism, and set the required bit masking when approximation is enabled. The need for approximations could be due to a variety of reasons. Architectural choice, external and run-time factors, nature of the algorithm, quick iterations for initial results, and power savings due to approximations are only some of these. Moreover, the impact of approximations may depend on the nature of input data, as can be seen from the experimental results in Sect. 7.3.2. Thus, by providing a compact set of software routines all these scenarios are addressed. Note that the compiler can also be optimized to automatically discover opportunities for approximation [CMR13, BC10]. Adding these capabilities to the ProACt compiler is left for future work. In the current state of the ProACt compiler framework, *when* and *where* to approximate, and the *granularity* of approximations are left to the programmer to decide.

In the next section, the experimental evaluation of ProACt is provided.

7.3   ProACt Evaluation

In this section, the experimental evaluation of ProACt is presented. First, a brief overview of the ProACt FPGA implementation is given. The experimental results for different applications using ProACt hardware are presented afterwards.

---

[6]https://www.gnu.org/s/binutils/.

[7]https://sourceware.org/newlib/.

7.3.1   FPGA Implementation Details

In order to study the hardware characteristics and the feasibility of the concept, a ProACt FPGA prototype is built using a Xilinx Zynq FPGA. This prototype is the basis of all the subsequent evaluations. A fixed 128 entry approximation look-up table is used in this ProACt prototype FPGA board. The table size is set to 128 mainly to utilize the FPGA resources for hardware implementation efficiently. In general, as the table size increases, the hit-rate and thereby the speed-up resulting from approximations increases. The general impact has already been observed by others, see, e.g., [ACV05, CFR98]. A number of design decisions are taken in the prototype for simplicity. The look-up table is unique and does not take care of the context switching of software threads in a multi-process and multi-threaded OS environment. Thus, the float operations from different threads feed into the same look-up table and consequently are treated alike. When multiple software threads are working on the same image context, this in fact is to some extent advantageous for the approximations due to spatial locality of the data. However, if the software threads execute vastly different programs, this aspect could be disadvantageous too. The thread level safety for approximations is left to the supervisor program (typically an OS) and a rapid switching mechanism (enable, disable, or change the approximation level) is provided with the SXL instruction. The current version of the ProACt compiler does not automatically discover opportunities for approximation. Hence, in this evaluation setup the programmer identifies such scenarios and writes the application utilizing on-demand approximation based on the ProACt system libraries.

The ProACt application development targeting the Zynq FPGA is shown in Fig. 7.3.

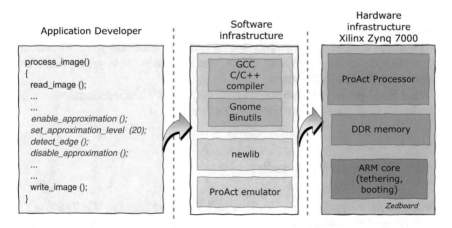

Fig. 7.3 Application development for ProACt Xilinx Zynq hardware

Table 7.1 ProACt FPGA hardware prototype details

Frequency:	100 MHz*	FPGA: Xilinx Zynq-XC7Z020
LUTs:	42,527	Prototype board: Digilent ZedBoard
Registers:	33,049	FPGA Tools: Xilinx Vivado 2016.2
Power/MHz:	18.66 mW/MHz	

*Clock frequency of 100 MHz as supported by the prototype Zedboard

The application developer writes the software in C/C++. The ProACt GCC cross-compiler compiles this application targeting newlib. The cross-compiler compiles the application in a host computer. Initial debugging and profiling is carried out using other GCC utilities. The cycle accurate ProACt emulator is used to emulate the system and make sure that the timing requirements are met. Note that the capability and scope of this emulator is limited due to the huge volume of information to be processed (even a simple *HelloWorld* C program can span thousands of cycles) and limited support for the low level system calls. Further, the application is transferred to the Xilinx Zynq development board and then run in the FPGA hardware.

The hardware details of the evaluation prototype are given in Table 7.1. The processor working frequency of 100 MHz is set by the clock in the prototype board.[8]

A useful *figure-of-merit* for the comparison of the prototype with other designs is the total power consumption per MHz. ProACt takes about 18.66 mW/MHz. The achieved value is easily comparable with other open source 64-bit processors such as OpenSPARC S1 targeted for FPGAs [JLG+14, Sun08]. OpenSPARC S1 has a total power dissipation of 965 mW at a F_{max} 65.77 MHz [JLG+14]. However, it must be emphasized that the prototype board, processor architecture, implementation details, and the I/O peripherals are widely different between these processors. Besides, current stage of the ProACt prototype is only a *proof-of-concept* of our methodology. Future research needs to consider more efficient look-up table architectures such as [IABV15] that can potentially improve the caching and retrieving mechanism. Further, similar hardware schemes and low power techniques like [GAFW07, AN04] are necessary for targeting ProACt to power critical embedded systems.

This FPGA prototype implementation is the basis of all experimental results and benchmarks presented in the subsequent sections.

7.3.2   Experimental Results

Two different categories of applications are used to evaluate ProACt. The first one is an image processing application and the second set consists of mathematical

---

[8]http://zedboard.org/product/zedboard.

functions from scientific computing. These experiments evaluate the performance of ProACt and also tests the on-demand approximation switching feature. All applications are written in C language, compiled using ProACt GCC compiler, and executed in the ProACt FPGA hardware prototype. All programs have a computationally expensive *core algorithm* which is the focus of this evaluation. A top level supervisor program controls the approximations and invokes the core algorithm. Thus, the same algorithm is run with different approximation schemes set by the supervisor program. Further, the approximation behavior of only the core algorithm is modified by the supervisor program.

The results are further analyzed in the host computer. The scope of approximations in this experimental evaluation is restricted to floating point division only. It has to be noted that the scheme used in these experiments (approximation control by a supervisor program) is only for evaluation purpose, and in other implementations the core-algorithm can also control the approximations. In the following, the experiments performed in the two categories are discussed:

7.3.2.1   Image Edge Detection

Table 7.2 shows the results from a case study on edge detection using ProACt. Here, the core algorithm is the edge detection routine. Image processing applications are very suitable for approximations since there is an inherent human perceptual limitation in processing image information. This case study uses a contrast-based edge detection algorithm [Joh90]. A threshold function is also used to improve the contrast of the detected edges. The minor differences in the pixel values are rounded off to the nearest value in this post-processing stage. This is another important aspect in reducing the differences introduced due to approximations.

The top row of images (Set 1) in Table 7.2 are generated with approximations disabled by the supervisory program. The middle row (Set 2) is generated with approximations enabled, and the last row shows bar plots for the hardware cycles taken by the core algorithm, along with the speed-up obtained. As evident from Table 7.2, ProACt is able to generate images with negligible loss of quality with performance improvements averaging more than 25%. Furthermore, the speed-up is much higher in images with more uniform pixels, as evident from the second image, *IEEE-754* (35% faster). This has to be expected since such a sparse input data set has higher chances of computation reuse.

7.3.2.2   Scientific Functions

ProACt is also evaluated with several scientific functions as the core algorithm. These scientific functions use floating point operations and are computationally expensive. A subset of the results is given in Table 7.3. The first row (non-shaded) in each set is the high accuracy version of the algorithms, i.e., with hardware approximations disabled. The second row (shaded) shows the result with

Table 7.2 Edge detection with approximations

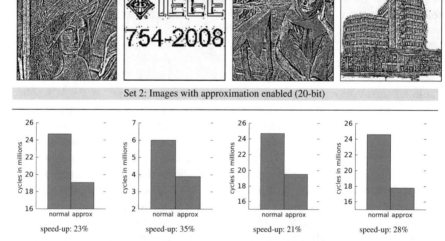

Lena	IEEE-754	Barbara	Building

Set 1: Images with approximation disabled *(reference)*

Set 2: Images with approximation enabled (20-bit)

Hardware cycles taken and speed-up with approximations

speed-up: 23%   speed-up: 35%   speed-up: 21%   speed-up: 28%

Images generated from ProACt FPGA hardware

Set 1 reference images are with normal processing

Set 2 images are with approximation enabled (20-bit)

approximations turned on. The absolute value of the deviation of the results ($|\Delta y|$) with approximation from the high accurate version is given in the third column, along with the respective speed-up obtained in fourth column (column d). We have run the experiments repeatedly 100 times with random inputs. The results shown are the average of the numbers obtained. The accuracy loss ($|\Delta y|$, given in column c), is only in the 4th decimal place, or lower in all the experiments.

The speed-up (column d) and the accuracy loss (column c) in Table 7.3 shows that on-demand approximations can significantly reduce the computation load with an acceptable loss in accuracy. Functions such as cosh and tanh can be approximated very well with a speed-up of more than 30% with an accuracy loss in the range of 0.0001.

Table 7.3 Math functions with approximations

appx Level (a)	Cycles n (b)	$\|\Delta y\|$ x10⁻³ (c)	Speed up% (d)	appx Level (a)	Cycles n (b)	$\|\Delta y\|$ x10⁻³ (c)	Speed up% (d)
$y = \sinh(x)$				$y = \sinh^{-1}(x)$			
−1	11,083	0.00	0.00	−1	76,899	0.00	0.00
20	7791	0.15	29.70	20	72,506	3.91	5.71
$y = \cosh(x)$				$y = \cosh^{-1}(x)$			
−1	10,820	0.00	0.00	−1	78,616	0.00	0.00
20	7501	0.14	30.67	20	73,843	2.14	6.07
$y = \tanh(x)$				$y = \tanh^{-1}(x)$			
−1	10,848	0.00	0.00	−1	7698	0.00	0.00
20	7505	0.10	30.82	20	6135	0.93	20.30
$y = j1(x)$ (Bessel series I)				$y = y1(x)$ (Bessel series II)			
−1	16,065	0.00	0.00	−1	16,941	0.00	0.00
20	14,432	0.01	10.16	20	14,445	0.01	14.73

Functions $y = f(x)$ evaluated in ProACt FPGA hardware
Results averaged over 100 random input values of x
Shaded rows are results with approximations enabled
(a) Approximation level (set with SXL instruction)
    −1: high accurate result (approximation fully disabled)
    20: 20-bit approximation in float division.
(b) Number of machine cycles (n) taken for computation.
(c) Accuracy, $\|\Delta y\| = |y_{-1} - y_{20}| \times 10^{-3}$

(d) Speed-up from approximation $= \dfrac{n_{-1} - n_{20}}{n_{-1}} \times 100\ \%$

Notes:
sinh, cosh, tanh computed as Taylor series expansion.
$\sinh^{-1}(x) = ln(x + \sqrt{x^2 + 1})$, $\cosh^{-1}(x) = ln(x + \sqrt{x^2 - 1})$
$\tanh^{-1}(x) = 0.5 * ln\left(\dfrac{1+x}{1-x}\right)$, $(j1, y1)$ computed with [Mos89]

7.3.2.3 Discussion

There is a substantial reduction in run time and machine cycles in ProACt with approximations enabled. About 25% speed-up, on an average, is obtained with approximations in the image edge detection applications shown in Table 7.2. This is also reflected in Table 7.3, where some of the mathematical functions are evaluated more than 30% faster. The dynamic power consumption of the system also decreases when approximations are enabled, since this speed-up directly corresponds to a reduction in the whole hardware activity.

The throughput and performance of a system, taken as a whole, is largely governed by Amdahl's law [Amd07]. I/O read and write are the main performance bottlenecks in the ProACt FPGA prototype. Consequently all the algorithms have been implemented to read the input data all at once, process, and then write out the result in a subsequent step. To make a fair comparison, the costly I/O read-write steps are not accounted in the reported speed-up.

It is also worth mentioning that all experimental results presented in Tables 7.2 and 7.3 are calculated with the software, which is compiled with the GCC flag -ffast-math [Gcc16]. This flag enables multiple compiler level optimizations such as constant folding for floating point computations, rather than offloading every operation to the FPU hardware. Thus, it potentially optimizes the floating point operations while generating the software binary itself, and the speed-up due to ProACt approximation adds on top of that. In practice, the use of this compile time flag is application dependent since it does not preserve strict IEEE compliance.

7.4   Concluding Remarks

In this chapter, we presented the ProACt processor architecture for cross-layer approximate computing. ProACt is FPGA proven and comes with a complete open-source development framework. Further, we have demonstrated the advantages of ProACt using image processing and scientific computing programs. These experimental evaluations show that up to 30% performance improvement can be achieved with dynamic approximation control.

We conclude the ProACt processor architecture for cross-layer approximate computing here. The next chapter summarizes the important conclusions and future outlook of this book.

Chapter 8
Conclusions and Outlook

In this book, algorithms and methodologies for the approximate computing paradigm have been proposed. Approximate computing hinges on cleverly using controlled inaccuracies (errors) in the operation for performance improvement. The key idea is to trade off correct computation against energy or performance. Approximate computing can address the growing demands of computational power for the current and future systems. Applications such as multi-media processing and compressing, voice recognition, web search, or deep learning are just a few examples where this novel computational paradigm provides huge benefits. However, this technology is still in its infancy and not widely adopted to mainstream. This is because of the lack of efficient design automation tools needed for approximate computing. This book provides several novel algorithms for the design automation of the approximation circuits. Our methodologies are efficient, scalable and significantly advance the current state-of-the-art of the approximate hardware design. We have addressed the important facets of approximate computing—from formal verification and error guarantees to synthesis and test of approximation systems.

Each chapter in this book presented one main contribution towards the realization of an approximate computing system. The first two chapters on verification explain the algorithms for formally verifying a system in the presence of functional errors. The algorithms and methodologies explained in these chapters can determine and prove the limits of approximation errors, both in combinational and sequential systems. The existing techniques based on statistical methods are inadequate to comprehensively verify the error bounds of such circuits. Our approaches provide the solution to a crucial and much needed approximation verification problem— *guarantees* on the bounds of the errors committed.

Further, automated synthesis approaches that can optimize the design with error guarantees have been presented in Chap. 5. The existing approximation synthesis techniques have several shortcomings (see Sect. 5.1 in Chap. 5 for details). In comparison to these approaches, our techniques can address the requirements on

© Springer Nature Switzerland AG 2019
A. Chandrasekharan et al., *Design Automation Techniques for Approximation Circuits*, https://doi.org/10.1007/978-3-319-98965-5_8

several error metrics that are specified together, and provide a formal guarantee on the error limits. The algorithms presented in Chap. 5 especially those based on AIGs scale very well with the circuit size. Evaluation on a wide range of circuits shows that our methodology is often better and even provides new avenues for optimization when compared to hand- crafted architecturally approximated circuits. Automated synthesis with error guarantees is a must for adopting approximate computing on a wider scale.

The next chapter dealt with the post-production test for approximate computing. The approximation-aware test technique detailed has a significant potential for yield improvement. To the best of our knowledge, this technique is the first systematic approach developed that considers the impact of design level approximations in test. The introduced test methodology does not change the established test techniques radically. Hence, it is relatively straightforward to adopt our techniques to the existing test flow—a fact that can lower the adoption barrier significantly, considering that the test has to be taken care of in design and fabrication, and has a profound impact in the final yield of IC manufacturing.

The final Chap. 7 provided the details of an on-demand approximation micro-processor called ProACt. ProACt is a high performance processor architecture with 64-bit addressing L1, L2 cache memories and supports features such as DMA for high throughput. The processor can do dynamic hardware level approximations, controlled and monitored using software. Thus, ProACt is best suited for cross-layer approximate computing where the hardware and software work together to achieve superior performance through approximations.

All the algorithms and methodologies explained in this book have been implemented and thoroughly evaluated. Besides, the underlying principles have been demonstrated on a wide range of benchmarks and use cases. In particular, the techniques on approximation verification and synthesis are available publicly as part of the *aXc* software framework. The processor prototype ProACt is also available publicly.

8.1 Outlook

The algorithms and methodologies presented in this book will alleviate several hurdles to make approximate computing a mainstream technology. Nevertheless, important future directions in each of the main topics—verification, synthesis, test, and architecture—can be identified. Extending the approximation synthesis techniques to state-based systems is an important direction of future research. A major challenge to this problem is to develop scalable algorithms that can ensure the error bounds during the synthesis process. The scalability of sequential approximation verification methodologies explained in Chap. 4, in the context of synthesis, needs to be studied.

In the test domain, approximation aware diagnosis and circuit rectification techniques that typically follow a post-production test run have to be investigated. This can potentially improve the overall Engineering Change Order (ECO) and re-spin time. The basic principles of SAT-based diagnosis can be applied to the approximation circuits cf. [SVAV06, CMB08]. An approximation-aware problem formulation step could be all what is needed in such a scheme. Further, one potential area of research for approximation-aware circuit re-synthesis in the re-spin stages will be to use *Quantifiable Boolean Formula* (QBF) and the ECO spare cells [REF17].

Another avenue of research related to cross-layer approximate computing is in the domain of compiler optimizations. Developing adaptive self- learning systems using an enhanced ProACt compiler is an important future research. This may be developed similar to the *dynamic power knobs* reported in [HSC$^+$11]. On the ProACt hardware side, different cache architectures need to be investigated for improving the approximation look-up table [IABV15].

References

[A⁺16]   K. Asanovic et al., The rocket chip generator, in *Technical Report UCB/EECS-2016-17, EECS Department*, University of California, Berkeley, 2016

[ACV05]   C. Alvarez, J. Corbal, M. Valero, Fuzzy memoization for floating-point multimedia applications. IEEE Trans. Comput. **54**, 922–927 (2005)

[AGM14]   L. Amarú, P.E. Gaillardon, G. De Micheli, Majority-inverter graph: a novel data-structure and algorithms for efficient logic optimization, in *Design Automation Conference*, vol. 194, pp. 1–194:6 (2014)

[AGM15]   L. Amarù, P.E. Gaillardon, G. De Micheli, The EPFL combinational benchmark suite, in *International Workshop on Logic Synthesis* (2015)

[Amd07]   G.M. Amdahl, Validity of the single processor approach to achieving large scale computing capabilities. IEEE Solid State Circuits Soc. Newsl. **12**, 19–20 (2007). Reprinted from the AFIPS conference

[AN04]   J.H. Anderson, F.N. Najm. Power estimation techniques for FPGAs. IEEE Trans. Very Large Scale Integration Syst. **12**, 1015–1027 (2004)

[And99]   H. Andersen, An introduction to binary decision diagrams, in *Lecture notes for Efficient Algorithms and Programs, The IT University of Copenhagen*, 1999

[Aok16]   Aoki Laboratory – Graduate School of Information Sciences. Tohoku University, 2016

[BA02]   M.L. Bushnell, V. Agrawal, *Essentials of Electronic Testing for Digital, Memory and Mixed-Signal VLSI Circuits* (Springer, Boston, 2002)

[BB04]   P. Bjesse, A. Boralv, Dag-aware circuit compression for formal verification. *International Conference on Computer Aided Design*, pp. 42–49 (2004)

[BC10]   W. Baek, T.M. Chilimbi, Green: a framework for supporting energy-conscious programming using controlled approximation, in *ACM SIGPLAN Notices*, vol. 45, pp. 198–209 (2010)

[BC14]   A. Bernasconi, V. Ciriani, 2-SPP approximate synthesis for error tolerant applications, in *EUROMICRO Symposium on Digital System Design*, pp. 411–418 (2014)

[BCCZ99]   A. Biere, A. Cimatti, E. Clarke, Y. ZhuH, Symbolic model checking without BDDs, in *Tools and Algorithms for the Construction and Analysis of Systems*, pp. 193–207 (1999)

[BDES14]   B. Becker, R. Drechsler, S. Eggersglüß, M. Sauer, Recent advances in SAT-based ATPG: non-standard fault models, multi constraints and optimization, in *International Conference on Design and Technology of Integrated Systems in Nanoscale Era*, pp. 1–10 (2014)

© Springer Nature Switzerland AG 2019
A. Chandrasekharan et al., *Design Automation Techniques
for Approximation Circuits*, https://doi.org/10.1007/978-3-319-98965-5

[BHR+12]   J. Bachrach, V. Huy, B. Richards, Y. Lee, A. Waterman, R. Avizienis, J. Wawrzynek, K. Asanovic, Chisel: constructing hardware in a Scala embedded language, in *Design Automation Conference*, pp. 1212–1221 (2012)

[BHvMW09]   A. Biere, R. Heule, H. van Maaren, T. Walsh, *Handbook of Satisfiability* (IOS Press, Berlin, 2009)

[Bra13]   A.R. Bradley, Incremental, inductive model checking, in *International Symposium on Temporal Representation and Reasoning*, pp. 5–6 (2013)

[Bre04]   M.A. Breuer, Determining error rate in error tolerant VLSI chips, in *Electronic Design, Test and Applications*, pp. 321–326 (2004)

[Bro90]   F.M. Brown, *Boolean Reasoning: The Logic of Boolean Equations* (Kluwer, Boston, 1990)

[Bry86]   R.E. Bryant, Graph-based algorithms for Boolean function manipulation. IEEE Trans. Comput. **35**, 677–691 (1986)

[Bry95]   R.E. Bryant, Binary decision diagrams and beyond: enabling techniques for formal verification, in *International Conference on Computer Aided Design*, pp. 236–243 (1995)

[BW96]   B. Bollig, I. Wegener, Improving the variable ordering of OBDDs is NP-complete. IEEE Trans. Comput. **45**, 993–1002 (1996)

[CCRR13]   V.K. Chippa, S.T. Chakradhar, K. Roy, A. Raghunathan, Analysis and characterization of inherent application resilience for approximate computing, in *Design Automation Conference*, pp. 1–9 (2013)

[CD96]   J. Cong, Y. Ding, Combinational logic synthesis for LUT based field programmable gate arrays. ACM Trans. Des. Autom. Electron. Syst. **1**, 145–204 (1996)

[CDA+12]   J. Constantin, A. Dogan, O. Andersson, P. Meinerzhagen, J.N. Rodrigues, D. Atienza, A. Burg, TamaRISC-CS: an ultra-low-power application-specific processor for compressed sensing, in *VLSI of System-on-Chip*, pp. 159–164 (2012)

[CEGD18]   A. Chandrasekharan, S. Eggersglüß, D. Große, R. Drechsler, Approximation-aware testing for approximate circuits, in *ASP Design Automation Conference*, pp. 239–244 (2018)

[CFR98]   D. Citron, D. Feitelson, L. Rudolph, Accelerating multi-media processing by implementing memoing in multiplication and division units, in *International Conference on Architectural Support for Programming Languages and Operating Systems*, pp. 252–261 (1998)

[CGD17a]   A. Chandrasekharan, D. Große, R. Drechsler, ProACt: a processor for high performance on-demand approximate computing, in *ACM Great Lakes Symposium on VLSI*, pp. 463–466 (2017)

[CGD17b]   A. Chandrasekharan, D. Große, R. Drechsler, Yise – a novel framework for boolean networks using Y-inverter graphs, in *International Conference on Formal Methods and Models for Codesign*, pp. 114–117 (2017)

[CMB05]   K.H. Chang, I.L. Markov, V. Bertacco, Post-placement rewiring and rebuffering by exhaustive search for functional symmetries, in *International Conference on Computer Aided Design*, pp. 56–63 (2005)

[CMB08]   K.H. Chang, I.L. Markov, V. Bertacco, Fixing design errors with counterexamples and resynthesis. IEEE Trans. Comput. Aided Des. Circuits Syst. **27**, 184–188 (2008)

[CMR13]   M. Carbin, S. Misailovic, M.C. Rinard, Verifying quantitative reliability for programs that execute on unreliable hardware, in *International Conference on Object-Oriented Programming Systems, Languages, and Applications*, pp. 33–52 (2013)

[CMV16]   S. Chakraborty, K.S. Meel, M.Y. Vardi, Algorithmic improvements in approximate counting for probabilistic inference: from linear to logarithmic SAT calls, in *International Joint Conference on Artificial Intelligence*, pp. 3569–3576 (2016)

[Coo71]   S.A. Cook, The complexity of theorem-proving procedures, in *Proceedings of the Third Annual ACM Symposium on Theory of Computing*, pp. 151–158 (1971)

[CSGD16a]   A. Chandrasekharan, M. Soeken, D. Große, R. Drechsler, Approximation-aware rewriting of AIGs for error tolerant applications, in *International Conference on Computer Aided Design*, pp. 83:1–83:8 (2016)

[CSGD16b]   A. Chandrasekharan, M. Soeken, D. Große, R. Drechsler, Precise error determination of approximated components in sequential circuits with model checking, in *Design Automation Conference*, pp. 129:1–129:6 (2016)

[CWK$^+$15] J. Constantin, L. Wang, G. Karakonstantis, A. Chattopadhyay, A. Burg, Exploiting dynamic timing margins in microprocessors for frequency-over-scaling with instruction-based clock adjustment, in *Design, Automation and Test in Europe*, pp. 381–386 (2015)

[DB98]   R. Drechsler, B. Becker, *Binary Decision Diagrams: Theory and Implementation* (Springer, New York, 1998)

[Een07]   N. Een, Cut sweeping, in *Cadence Design Systems*, Technical Report, 2007

[EMB11]   N. Een, A. Mishchenko, R.K. Brayton, Efficient implementation of property directed reachability, in *International Conference on Formal Methods in CAD*, pp. 125–134 (2011)

[ESCB12]   H. Esmaeilzadeh, A. Sampson, L. Ceze, D. Burger, Architecture support for disciplined approximate programming. ACM SIGPLAN Not. **47**, 301–312 (2012)

[FS83]   H. Fujiwara, T. Shimono, On the acceleration of test generation algorithms. IEEE Trans. Comput. **32**, 1137–1144 (1983)

[GA15]   GeAr-ApproxAdderLib, Chair for Embedded Systems – Karlsruhe Institute of Technology, 2015

[GAFW07]   S. Gupta, J. Anderson, L. Farragher, Q. Wang, CAD techniques for power optimization in virtex-5 FPGAs, in *IEEE Custom Integrated Circuits Conference*, pp. 85–88 (2007)

[Gcc16]   GCC – the GNU Compiler Collection 6.1, 2016

[GMP$^+$11] V. Gupta, D. Mohapatra, S.P. Park, A. Raghunathan, K. Roy, IMPACT: imprecise adders for low-power approximate computing, in *International Symposium on Low Power Electronics and Design*, pp. 409–414 (2011)

[HABS14]   M.H. Haghbayan, B. Alizadeh, P. Behnam, S. Safari, Formal verification and debugging of array dividers with auto-correction mechanism, in *VLSI Design*, pp. 80–85 (2014)

[Hau96]   J.R. Hauser, Handling floating-point exceptions in numeric programs. ACM Trans. Program. Lang. Syst. **18**, 139–174 (1996)

[HP11]   J.L. Hennessy, D.A. Patterson, *Computer Organization and Design, Fourth Edition: The Hardware/Software Interface* (Morgan Kaufmann, Waltham, 2011)

[HS02]   G.D. Hachtel, F. Somenzi, *Logic Synthesis and Verification Algorithms* (Kluwer, Boston, 2002)

[HSC$^+$11] H. Hoffmann, S. Sidiroglou, M. Carbin, S. Misailovic, A. Agarwal, M. Rinard, Dynamic knobs for responsive power-aware computing. ACM SIGPLAN Not. **46**, 199–212 (2011)

[HYH99]   M.C. Hansen, H. Yalcin, J.P. Hayes, Unveiling the ISCAS-85 benchmarks: a case study in reverse engineering. IEEE Des. Test **16**, 72–80 (1999)

[IABV15]   Z. Istvan, G. Alonso, M. Blott, K. Vissers, A hash table for line-rate data processing. ACM Trans. Reconfig. Technol. Syst. **8**, 13:1–13:15 (2015)

[IS75]   O.H. Ibarra, S.K. Sahni, Polynomially complete fault detection problems. IEEE Trans. Comput. **C-24**, 242–249 (1975)

[ISYI09]   H. Ichihara, K. Sutoh, Y. Yoshikawa, T. Inoue, A practical approach to threshold test generation for error tolerant circuits, in *Asian Test Symposium*, pp. 171–176 (2009)

[Itu16]   International Telecommunication Union, 2016

126   References

[JLG+14]   R. Jia, C.Y. Lin, Z. Guo, R. Chen, F. Wang, T. Gao, H. Yang, A survey
             of open source processors for FPGAs, in *International Conference on Field
             Programmable Logic and Applications*, pp. 1–6 (2014)
[Joh90]   R.P. Johnson, Contrast based edge detection. J. Pattern Recogn. **23**, 311–318
             (1990). Elsevier Science.
[KGAKP14]   M. Kamal, A. Ghasemazar, A. Afzali-Kusha, M. Pedram, Improving efficiency
             of extensible processors by using approximate custom instructions, in *Design,
             Automation and Test in Europe*, pp. 1–4 (2014)
[KGE11]   P. Kulkarni, P. Gupta, M. Ercegovac, Trading accuracy for power with an
             underdesigned multiplier architecture, in *VLSI Design*, pp. 346–351 (2011)
[KK12]   A.B. Kahng, S. Kang, Accuracy-configurable adder for approximate arithmetic
             designs, in *Design Automation Conference*, pp. 820–825 (2012)
[Knu11]   D.E. Knuth, *The Art of Computer Programming*, vol. 4A (Addison-Wesley,
             Upper Saddle River, 2011)
[Knu16]   D.E. Knuth, *Pre-fascicle to The Art of Computer Programming, Section 7.2.2.
             Satisfiability*, vol. 4 (Addison-Wesley, Upper Saddle River, 2016)
[Kre88]   M.W. Krentel, The complexity of optimization problems. J. Comput. Syst. Sci.
             25, 743–755 (1988)
[KSS+14]   G. Karakonstantis, A. Sankaranarayanan, M.M. Sabry, D. Atienza, A. Burg, A
             quality-scalable and energy-efficient approach for spectral analysis of heart rate
             variability, in *Design, Automation and Test in Europe*, pp. 1–6 (2014)
[Lar92]   T. Larrabee, Test pattern generation using Boolean satisfiability. IEEE Trans.
             Comput. Aided Des. Circuits Syst. **11**, 4–15 (1992)
[LD11]   N. Li, E. Dubrova, AIG rewriting using 5-input cuts, in *International Conference
             on Computer Design*, pp. 429–430 (2011)
[LEN+11]   A. Lingamneni, C. Enz, J.L. Nagel, K. Palem, C. Piguet, Energy parsimonious
             circuit design through probabilistic pruning, in *Design, Automation and Test in
             Europe*, pp. 1–6 (2011)
[LHB05]   K.J. Lee, T.Y. Hsieh, M.A. Breuer, A novel test methodology based on error-rate
             to support error-tolerance, in *International Test Conference*, pp. 1–9 (2005)
[LHB12]   K.J. Lee, T.Y. Hsieh, M.A. Breuer, Efficient overdetection elimination of
             acceptable faults for yield improvement. IEEE Trans. Comput. Aided Des.
             Circuits Syst. **31**, 754–764 (2012)
[Lun16]   D. Lundgren, OpenCore JPEG Encoder – OpenCores community, 2016
[MCB06]   A. Mishchenko, S. Chatterjee, R.K. Brayton, Dag-aware aig rewriting a fresh
             look at combinational logic synthesis, in *Design Automation Conference*,
             pp. 532–535 (2006)
[MCBJ08]   A. Mishchenko, M. Case, R.K. Brayton, S. Jang, Scalable and scalably-verifiable
             sequential synthesis, in *International Conference on Computer Aided Design*,
             pp. 234–241 (2008)
[MHGO12]   J. Miao, K. He, A. Gerstlauer, M. Orshansky, Modeling and synthesis of quality-
             energy optimal approximate adders, in *International Conference on Computer
             Aided Design*, pp. 728–735 (2012)
[MMZ+01]   M.W. Moskewicz, C.F. Madigan, Y. Zhao, L. Zhang, S. Malik, Chaff: engi-
             neering an efficient SAT solver, in *Design Automation Conference*, pp. 530–535
             (2001)
[Mos89]   S.L.B. Moshier, *Methods and Programs for Mathematical Functions* (Ellis
             Horwood, Chichester, 1989)
[MZS+06]   A. Mishchenko, J.S. Zhang, S. Sinha, J.R. Burch, R.K. Brayton, M.C. Jeske,
             Using simulation and satisfiability to compute flexibilities in boolean networks.
             IEEE Trans. Comput. Aided Des. Circuits Syst. **25**, 743–755 (2006)
[Nav11]   Z. Navabi, *Digital System Test and Testable Design* (Springer, New York, 2011)
[Oma16]   Texas instruments OMAP L-1x series processors, 2016

[PH08]   D.A. Patterson, J.L. Hennessy, *Computer Organization and Design, Fourth Edition: The Hardware/Software Interface* (Morgan Kaufmann, Waltham, 2008)

[PL98]   P. Pan, C. Lin, A new retiming-based technology mapping algorithm for LUT-based FPGAs, in *International Symposium on FPGAs*, pp. 35–42 (1998)

[PMS$^+$16] A. Petkovska, A. Mishchenko, M. Soeken, G. De Micheli, R.K. Brayton, P. Ienne, Fast generation of lexicographic satisfiable assignments: enabling canonicity in SAT-based applications, in *International Conference on Computer Aided Design*, pp. 1–8 (2016)

[REF17]   H. Riener, R. Ehlers, G. Fey, CEGAR-based EF synthesis of boolean functions with an application to circuit rectification, in *ASP Design Automation Conference*, pp. 251–256 (2017)

[Rot66]   J.P. Roth, Diagnosis of automata failures: a calculus and a method. IBM J. Res. Dev. **10**, 278–281 (1966)

[RRV$^+$14] A. Ranjan, A. Raha, S. Venkataramani, K. Roy, A. Raghunathan, ASLAN: synthesis of approximate sequential circuits, in *Design, Automation and Test in Europe*, pp. 1–6 (2014)

[RS95]   K. Ravi, F. Somenzi, High-density reachability analysis, in *International Conference on Computer Aided Design*, pp. 154–158 (1995)

[SA12]   S. Sindia, V.D. Agrawal, Tailoring tests for functional binning of integrated circuits, in *Asian Test Symposium*, pp. 95–100 (2012)

[SAHH15]   M. Shafique, W. Ahmad, R. Hafiz, J. Henkel, A low latency generic accuracy configurable adder, in *Design Automation Conference*, pp. 1–6 (2015)

[Sat16]   SAT-Race–2016, International Conference on Theory and Applications of Satisfiability Testing, 2016

[SE12]   R. Drechsler S. Eggersglüß, *High Quality Test Pattern Generation and Boolean Satisfiability* (Springer, Boston, 2012)

[SG10]   D. Shin, S.K. Gupta, Approximate logic synthesis for error tolerant applications, in *Design, Automation and Test in Europe*, pp. 957–960 (2010)

[SG11]   D. Shin, S.K. Gupta, A new circuit simplification method for error tolerant applications, in *Design, Automation and Test in Europe*, pp. 1–6 (2011)

[SGCD16]   M. Soeken, D. Große, A. Chandrasekharan, R. Drechsler, BDD minimization for approximate computing, in *ASP Design Automation Conference*, pp. 474–479 (2016)

[SLFP16]   X. Sui, A. Lenharth, D.S. Fussell, K. Pingali, Proactive control of approximate programs, in *International Conference on Architectural Support for Programming Languages and Operating Systems*, pp. 607–621 (2016)

[SLJ$^+$13] M. Samadi, J. Lee, D.A. Jamshidi, A. Hormati, S. Mahlke, Sage: self-tuning approximation for graphics engines, in *International Symposium on Microarchitecture*, pp. 13–24 (2013)

[Som99]   F. Somenzi, Binary decision diagrams, in *NATO Science Series F: Computer and Systems Sciences*, vol. 173, pp. 303–366 (1999)

[Sun08]   Sun Microsystems Inc, OpenSPARC T1 Microarchitecture Specification, 2008

[SVAV06]   A. Smith, A. Veneris, M.F. Ali, A. Viglas, Fault diagnosis and logic debugging using boolean satisfiability, in *IEEE Transactions on Computer Aided Design of Circuits and Systems*, vol. 24, pp. 1606–1621 (2006)

[Thu06]   M. Thurley, sharpSAT: counting models with advanced component caching and implicit BCP, in *Theory and Applications of Satisfiability Testing*, pp. 424–429 (2006)

[Tse68]   G. Tseitin, On the complexity of derivation in propositional calculus, in *Studies in Constructive Mathematics and Mathematical Logic*, vol. 2, pp. 115–125 (1968)

[VARR11]   R. Venkatesan, A. Agarwal, K. Roy, A. Raghunathan, MACACO: modeling and analysis of circuits for approximate computing, in *International Conference on Computer Aided Design*, pp. 667–673 (2011)

[VCC⁺13]   S. Venkataramani, V.K. Chippa, S.T. Chakradhar, K. Roy, A. Raghunathan, Quality programmable vector processors for approximate computing, in *International Symposium on Microarchitecture*, pp. 1–12 (2013)

[VCRR15]   S. Venkataramani, S.T. Chakradhar, K. Roy, A. Raghunathan, Approximate computing and the quest for computing efficiency, in *Design Automation Conference*, pp. 1–6 (2015)

[vHLP07]   F. van Harmelen, V. Lifschitz, B. Porter, *Handbook of Knowledge Representation* (Elsevier Science, San Diego, 2007)

[VSK⁺12]   S. Venkataramani, A. Sabne, V. Kozhikkottu, K. Roy, A. Raghunathan, Salsa: systematic logic synthesis of approximate circuits, in *Design Automation Conference*, pp. 796–801 (2012)

[WLPA11]   A. Waterman, Y. Lee, D.A. Patterson, K. Asanovic, The RISC-V instruction set manual, volume i: base user-level ISA, in *Technical Report UCB/EECS-2011-62, EECS Department*, University of California, Berkeley, 2011

[WTV⁺17]   I. Wali, M. Traiola, A. Virazel, P. Girard, M. Barbareschi, A. Bosio, Towards approximation during test of integrated circuits, in *IEEE Workshop on Design and Diagnostics of Electronic Circuits and Systems*, pp. 28–33 (2017)

[XMK16]   Q. Xu, T. Mytkowicz, N.S. Kim, Approximate computing: a survey. IEEE Des. Test **33**, 8–22 (2016)

[Yan91]   S. Yang, Logic synthesis and optimization benchmarks user guide version 3.0 (1991)

[YPS⁺15]   A. Yazdanbakhsh, J. Park, H. Sharma, P. Lotfi-Kamran, H. Esmaeilzadeh, Neural acceleration for GPU throughput processors, in *International Symposium on Microarchitecture*, pp. 482–493 (2015)

[YWY⁺13]   R. Ye, T. Wang, F. Yuan, R. Kumar, Q. Xu, On reconfiguration-oriented approximate adder design and its application, in *International Conference on Computer Aided Design*, pp. 48–54 (2013)

[ZGY09]   N. Zhu, W.L. Goh, K.S. Yeo, An enhanced low-power high-speed adder for error-tolerant application, in *International Symposium on IC Technologies, Systems and Applications*, pp. 69–72 (2009)

[ZPH04]   L. Zhang, M.R. Prasad, M.S. Hsiao, Incremental deductive inductive reasoning for SAT-based bounded model checking, in *International Conference on Computer Aided Design*, pp. 502–509 (2004)

Index

© Springer Nature Switzerland AG 2019
A. Chandrasekharan et al., *Design Automation Techniques
for Approximation Circuits*, https://doi.org/10.1007/978-3-319-98965-5

Printed in the United States
By Bookmasters